I0795635

IMAGES
of America

JIM THORPE (MAUCH CHUNK)

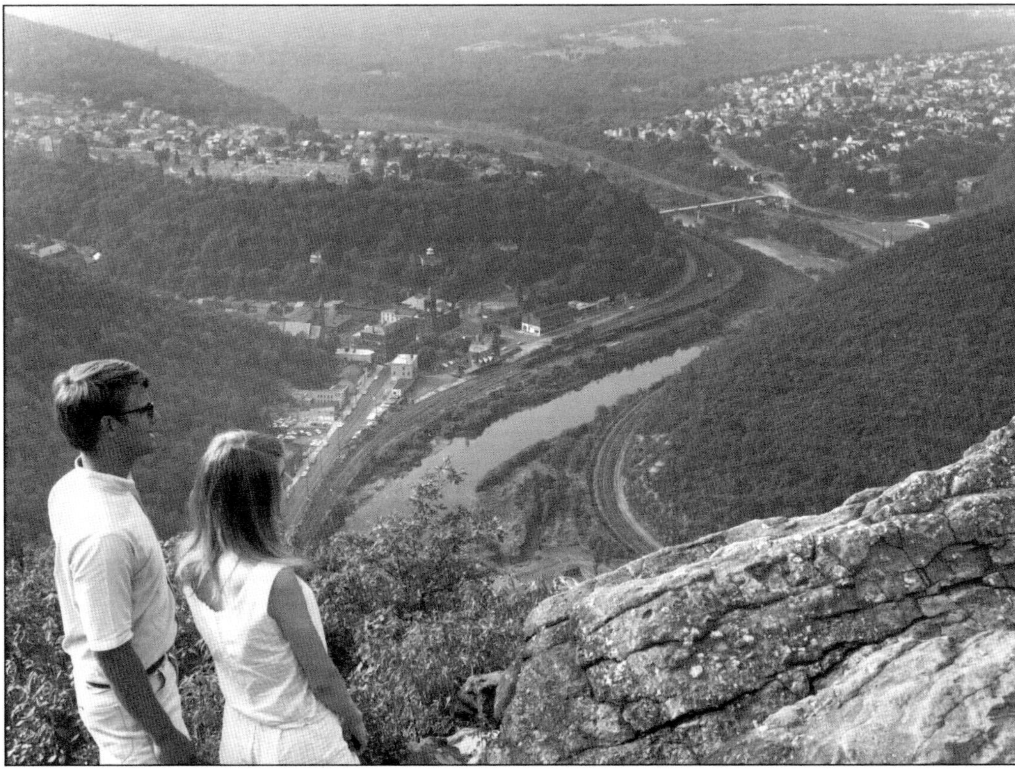

This view from Flagstaff Mountain, 1,000 feet above Jim Thorpe, shows the historic district of Jim Thorpe and the west bank of Lehigh River (left), connected to East Jim Thorpe (top right) by a bridge. The Heights section of town, including the old cemetery, overlooks the historic district. (Courtesy Agnes McCartney.)

The thirst for "black diamonds" has long diminished.
The hero-industrialists have been replaced by the hero-environmentalists.
The focus is on the sublime and picturesque.
And, yet, this old industrial outpost grants reminiscences of its glory days,
and we're astonished as we gaze at some of man's and nature's best creative offerings.
Heyday revisited indeed!
And, now, with clean air and shad in the river.

—John H. Drury

Cover photograph courtesy Beverly Sillers.

IMAGES
of America

JIM THORPE (MAUCH CHUNK)

John H. Drury and Joan Gilbert

ARCADIA
PUBLISHING

ISBN 978-0-7385-0963-1

Published by Arcadia Publishing
Charleston, South Carolina

Printed in the United States of America

Library of Congress Catalog Card Number: 2001093534

For all general information contact Arcadia Publishing at:
Telephone 843-853-2070
Fax 843-853-0044
E-mail sales@arcadiapublishing.com
For customer service and orders:
Toll-Free 1-888-313-2665

Visit us on the Internet at www.arcadiapublishing.com

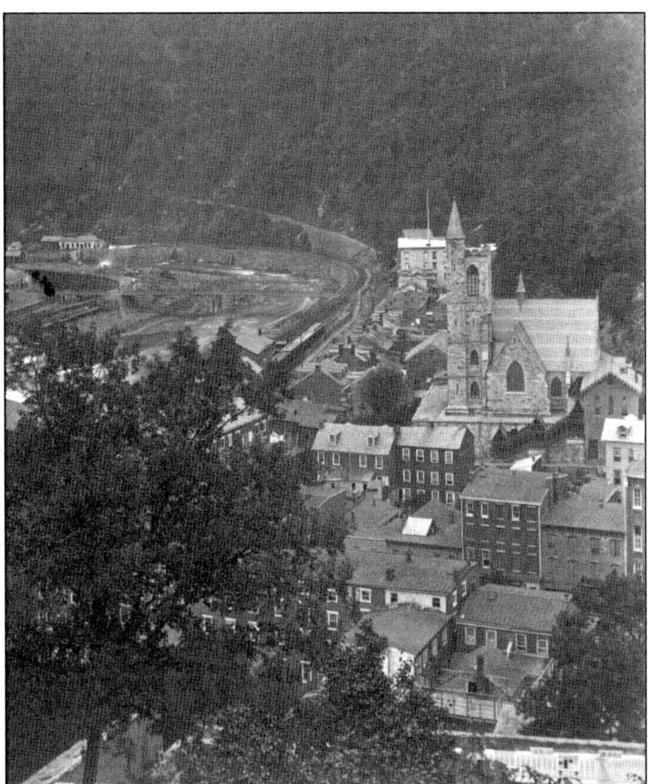

This view, taken from Mount Pisgah, looks over the rooftops to the historic district of Jim Thorpe. The road into town down Mansion House Hill hugs the base of Flagstaff Mountain.

Contents

ACKNOWLEDGMENTS

I offer my thanks to a multitude of families in town who shared their treasured photographs and memorabilia; Ray Holland, for his commitment to historic preservation and continuing generosity in sharing his extensive collections; Mary Wohnhaas, Mauch Chunk Museum and Cultural Center administrator, for organizing the collections of photographs; Vince Hydro, Jim Thorpe's foremost resident historian, for reviewing the manuscript; the late Laura Thomas, whose considerable energy, abilities, and love of community helped develop the Mauch Chunk Museum and Cultural Center; Joan Gilbert, for writing text and designing the layout of the award-winning Mauch Chunk Museum and Cultural Center; and Janet Drury, my helpmate throughout a lifetime of projects and causes.

—John H. Drury

Preface

As the proud Sac and Fox tribe member Jim Thorpe stood shyly before King Gustav of Sweden at the 1912 Stockholm Olympics to receive gold medals for his twin victories in the pentathlon and decathlon—a feat that has never been duplicated—the king proclaimed, "You, sir, are the greatest athlete in the world!"

Each small town has its own history, but few have a history more unique than Jim Thorpe, Pennsylvania. Beset by economic woes when the coal mines and railroads shut down, the town (previously known as Mauch Chunk) exchanged its name for the body and name of Jim Thorpe when the famous Olympian was entombed there in 1954.

Unlike those who rode wagon trains across the vast prairies to the American West, settling small towns along the way, most of Mauch Chunk's early settlers arrived on foot. They trudged along the old stagecoach road, traversing the mountain wilderness by way of the river valleys, lured by the promise of work in the developing coal industry. Some came from villages, towns, or cities in America. Most were immigrants—Poles, Slavs, Germans, English, Welsh, and Irish—fulfilling their dreams in a land where, it was rumored, opportunity and riches were theirs for the taking. As these men walked toward a new dawn, their hopes were as high as the mountains. When Josiah White and Erskine Hazard, Philadelphia business partners, established a coal-shipping depot (ultimately the Lehigh Coal & Navigation Company) at the confluence of the Lehigh River and Mauch Chunk Creek in 1818, the laborers named the mountain rising above the industrial outpost Mount Pisgah after the mountain that overlooked the biblical Promised Land. The 15,100-pound block of anthracite coal, glistening blue-black in the hub of the town now named Jim Thorpe, pays tribute to these men whose vision and hard labor helped to propel America into the industrial age.

This pictorial history follows the journey of the early settlers and those who came after them, beginning with the discovery and mining of black diamonds (coal), the early settlement, the scramble to build waterways and railroads to transport coal to eastern seaboard cities, and the amazing Switchback Gravity Railroad. The story continues through Mauch Chunk's golden age of industry and tourism, when railroads brought excursionists to the town in droves. The history finally addresses the town's decline and revival, as Mauch Chunk became Jim Thorpe.

Flooding was once Mauch Chunk's nemesis. In 1971, the Mauch Chunk Creek was dammed above town, creating a 2.5-mile-long, 325-acre lake. The scenic park encompasses part of the Switchback trail and offers facilities for camping, hiking, boating, fishing, swimming, and cross-country skiing. Today, tourism is on the rebound. In 2000, an estimated 300,000 visitors came to the quaint mountain town. The Lehigh Canal towpath, rail beds, and the glorious route of the Switchback Gravity Railroad have been recycled into hiking, mountain biking, and cross-country skiing trails. Where once the conch shells of boatmen sounded the presence of coal boats on the Lehigh Canal, rafters ride the white waters of the Lehigh River through its gorge. As mountain bikers pedal hard up Jim Thorpe's hilly streets, and tourists admire its Victorian buildings, visit its fine hostelries, take a train ride, or browse through its shops and museums, the glory days of yore are returning.

One

STRUGGLE

In 1791, anthracite coal, dubbed "stone coal" by derisive industrialists, was discovered in Summit Hill, Pennsylvania. The Lehigh Mine Company was founded and quarrying began. Coal was brought by horse and wagon nine miles over the mountains to the Lehigh River, and attempts were made to float it downstream by ark, or raft. Very few arks reached their destination. Most were wrecked in the steep, treacherous river. The venture was unprofitable, and the mine was abandoned.

Desperate for fuel when the British blockaded southern ports and put an embargo on bituminous coal during the War of 1812, Josiah White and Erskine Hazard, owners of a wire mill near Philadelphia, purchased a load of stone coal that survived the trip by ark down the Lehigh and Delaware Rivers. Disgusted at the coal's poor performance, a workman slammed the furnace door and left for the night. Several hours later, word came that the mill was on fire. When White and Hazard rushed to the scene, they found that the red glow was not the mill, but anthracite burning white-hot in the closed, ventilated furnace. By accident, they had discovered the secret to successfully burning stone coal.

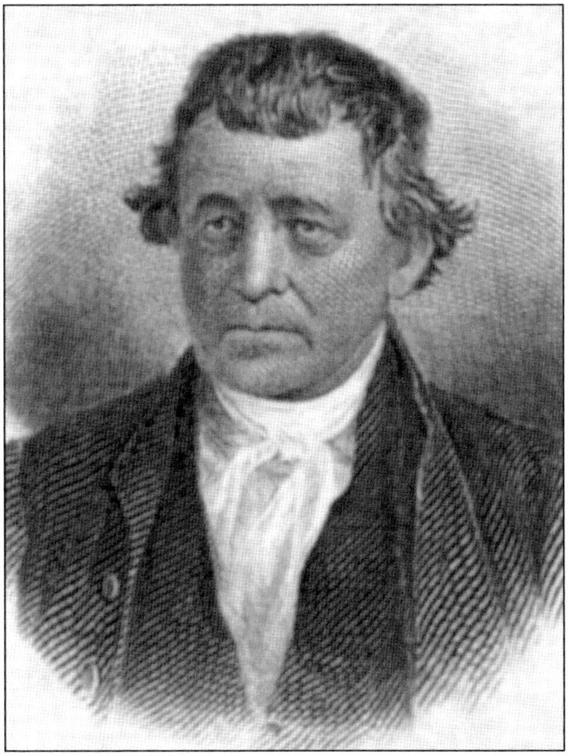

In 1817, Josiah White (1781–1850), shown here, set out to seek his fortune in the Pennsylvania anthracite fields. Short of cash, he asked business acquaintance Joseph Gillingham to lend him a horse. Gillingham refused. "If anything favorable results," declared White in his diary, "I will keep him clear of it." The loser was Gillingham. With his wire mill partner, Erskine Hazard, White leased the abandoned Summit Hill anthracite quarry and 10,000 acres of land for one ear of corn a year with the proviso that, by the third year, they ship a minimum of 40,000 bushels of anthracite coal to market. In 1822, Josiah White, Erskine Hazard, and George Hauto (later bought out for 10¢ on each ton of coal shipped) founded the Lehigh Coal & Navigation Company, the industrial giant whose coal mines, rail lines, and steamship lines would help to revolutionize American industry.

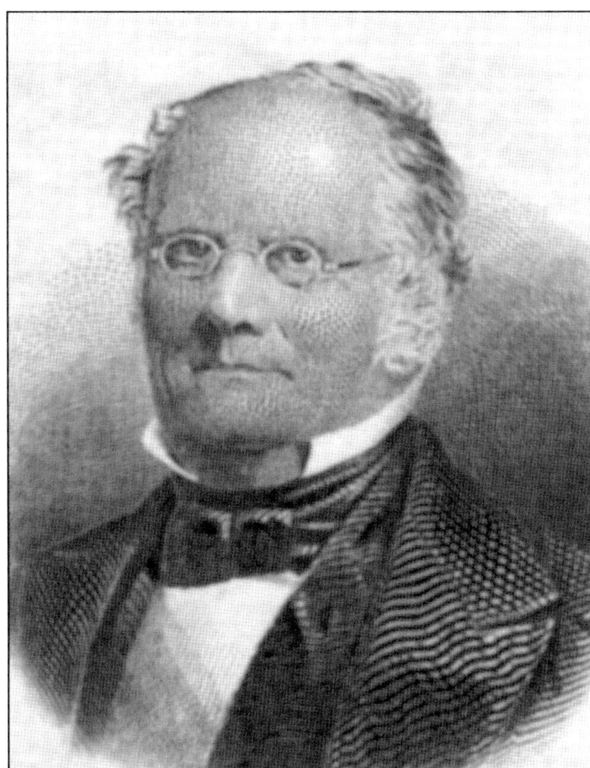

Josiah White was an engineering genius and visionary who, undaunted by environmental and financial challenges, founded the Lehigh Coal & Navigation Company. He also built the Lehigh Canal and the Switchback Gravity Railroad. Erskine Hazard, shown here, was White's worthy partner and trusted friend. White described Hazard's role in the Lehigh Coal & Navigation Company: "Erskine was the scribe and a good mechanic and excellent counsellor."

In 1818, a settlement grew up in the ravine near the coal-shipping outpost where the Mauch Chunk Creek enters the Lehigh River. The settlers named the village Mauch Chunk after the crouching mountain across the river, which the Lenape Indians called Mountain of the Sleeping Bear. This 1826 map was drawn by Solomon Roberts, Josiah White's nephew. (Courtesy Ray Holland.)

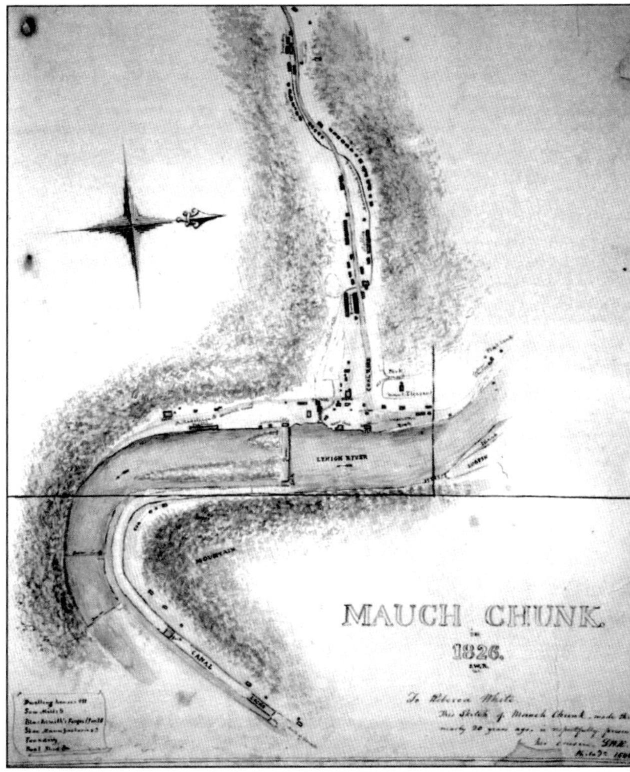

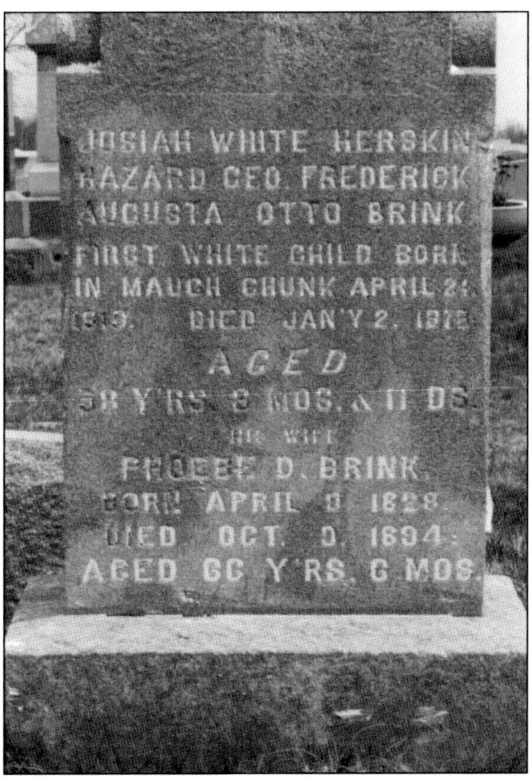

The first white child born in Mauch Chunk arrived on April 21, 1819, to the settlement's first woman, Margaret Brink, and her husband, Nicholas. The Brinks named their son after the town's three founders. With great ceremony, he was baptized Josiah White Erskine Hazard George F.A. Hauto Brink. In celebration, "the forest was illuminated with pine torches, plenty of old pure rye whiskey was drunk, and the noise and dancing was so great that it seemed as if the very tops of the pines had caught the infection, and kept time by swaying to and fro." —From Brink's tombstone, Summit Hill Cemetery.

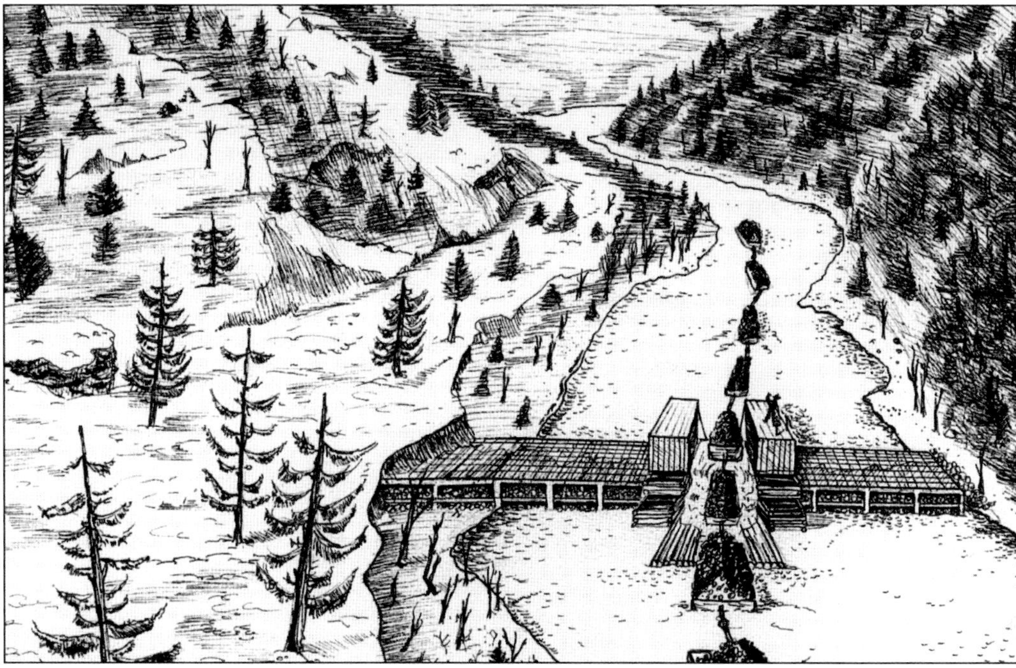

White and Hazard improved the coal road from the mine and secured navigation rights to the Lehigh River from the State of Pennsylvania. Descending one-way navigation (the shipment of coal downstream by ark) had to be operable by 1824. Two-way navigation (the construction of a canal) had to be operating by 1839. By 1820, coal arks were navigating the river by way of Josiah White's ingenious bear-trap lock and dam system. Built into dams, 14 bear-trap locks permitted trains of arks (as long as 180 feet) to pass downstream to Easton over the river shallows on artificially created floods.

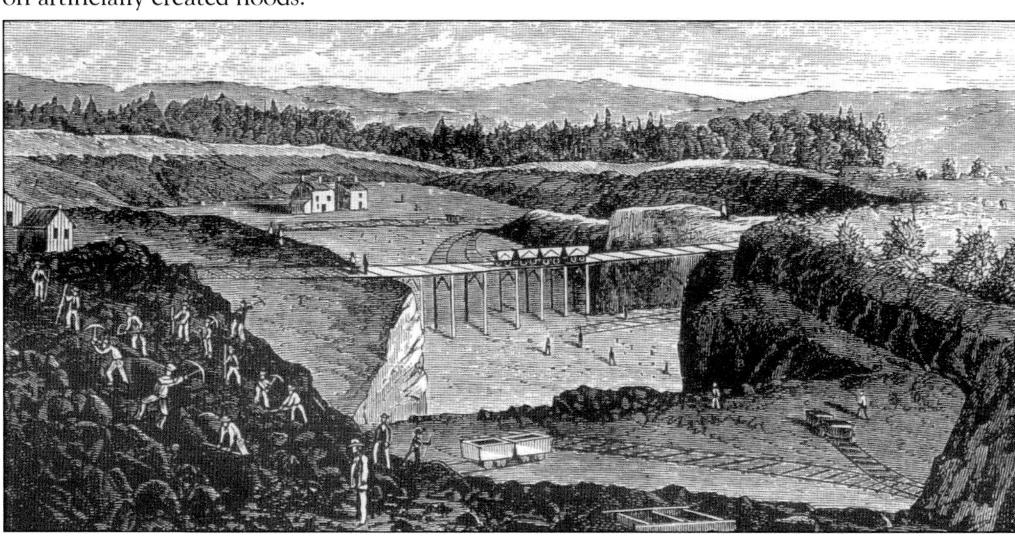

"Drove thro the cart way into the [Summit Hill] mine. Horse made a dead pause to give me time to get over my astonishment. A space of about four acres completely excavated. Walls of solid masses of glistening coal rising about 40 feet on every side. About two hundred men at work in every direction. A most active scene of bustle and business. The four sides and the bottoms one unvarying mass of coal!" —From the diary of Edmund S. Coxe, 1825.

In 1827, the Mauch Chunk Railroad was constructed on top of the coal road from the Summit Hill mine. Eight mules hauled a train of 14 empty coal cars up to the mine. They rode back to town in special cars and dined en route. With only a brakeman in control, the train made the downhill trip to Mauch Chunk in 30 minutes. Passengers rode the gravity railroad as early as 1829.

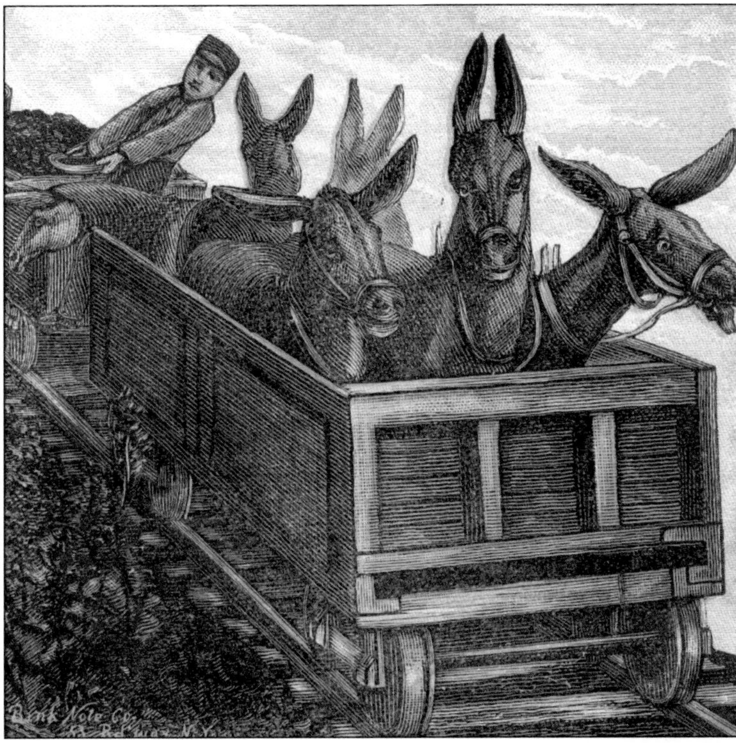

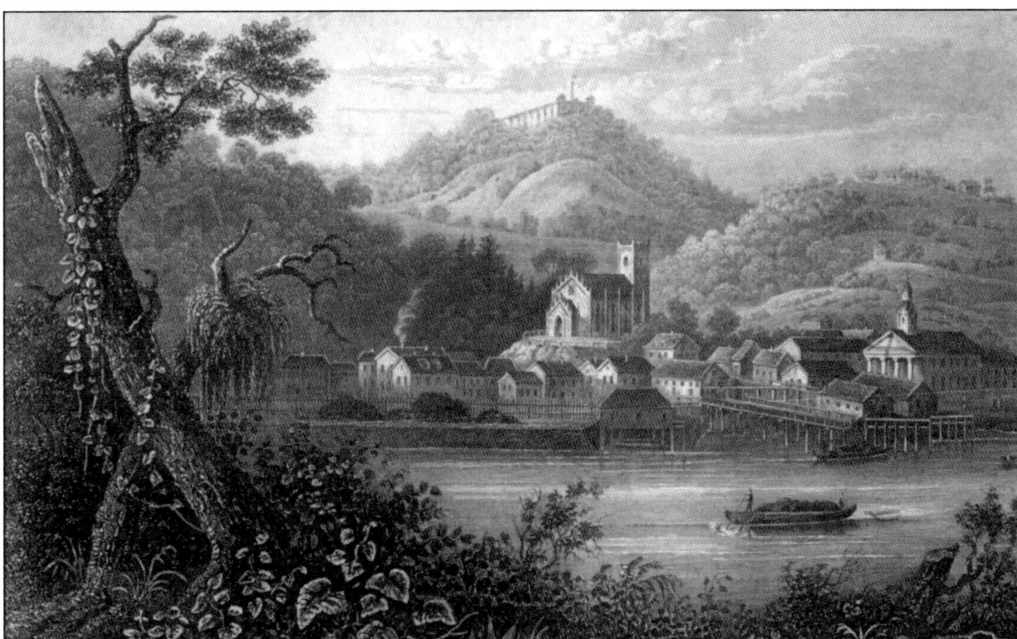

"The foot of the coal mountain rests upon the village of Mauch Chunk and you ascend to the mine, situated nine miles from the landing on top of the mountain, by the most noble and beautiful turnpike in the United States, and perhaps not surpassed in any country. Turnpike black with coal dust. Not very pleasant in a high wind. Clean faces not often to be met with on the road." —From the diary of Edmund S. Coxe, 1825.

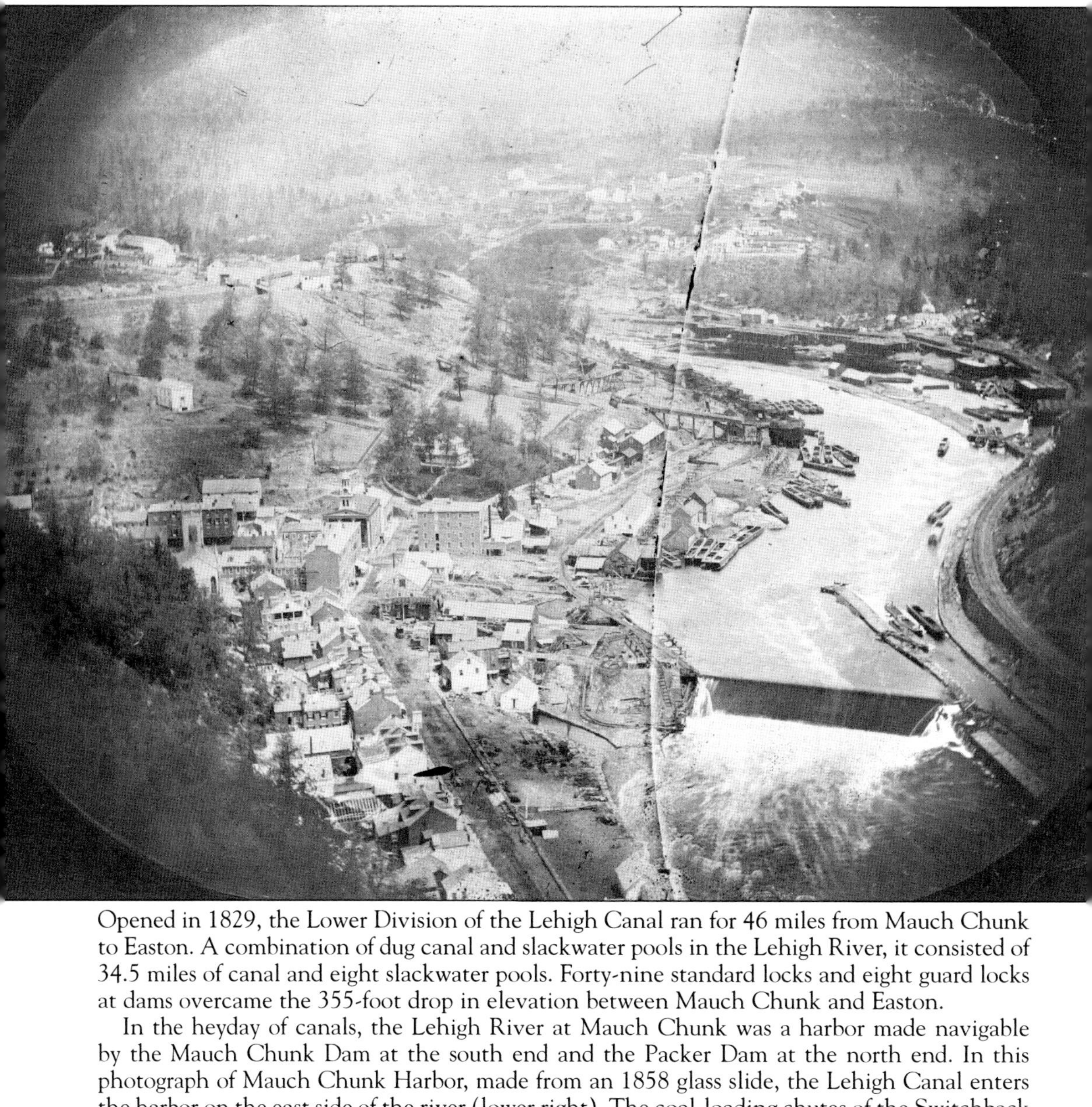

Opened in 1829, the Lower Division of the Lehigh Canal ran for 46 miles from Mauch Chunk to Easton. A combination of dug canal and slackwater pools in the Lehigh River, it consisted of 34.5 miles of canal and eight slackwater pools. Forty-nine standard locks and eight guard locks at dams overcame the 355-foot drop in elevation between Mauch Chunk and Easton.

In the heyday of canals, the Lehigh River at Mauch Chunk was a harbor made navigable by the Mauch Chunk Dam at the south end and the Packer Dam at the north end. In this photograph of Mauch Chunk Harbor, made from an 1858 glass slide, the Lehigh Canal enters the harbor on the east side of the river (lower right). The coal-loading chutes of the Switchback Gravity Railroad are on the west side of the river above the coal boats (center). Noticeably absent on the lower slopes of Mount Pisgah (left) are the mansion of Asa Packer (built in 1861) and the mansion of his son Harry (built in 1874). Also absent on the west bank of the river are the tracks of the Central Railroad of New Jersey, extended to Easton in 1867. The tracks of Asa Packer's Lehigh Valley Railroad (founded in 1855) may be seen on the east side.

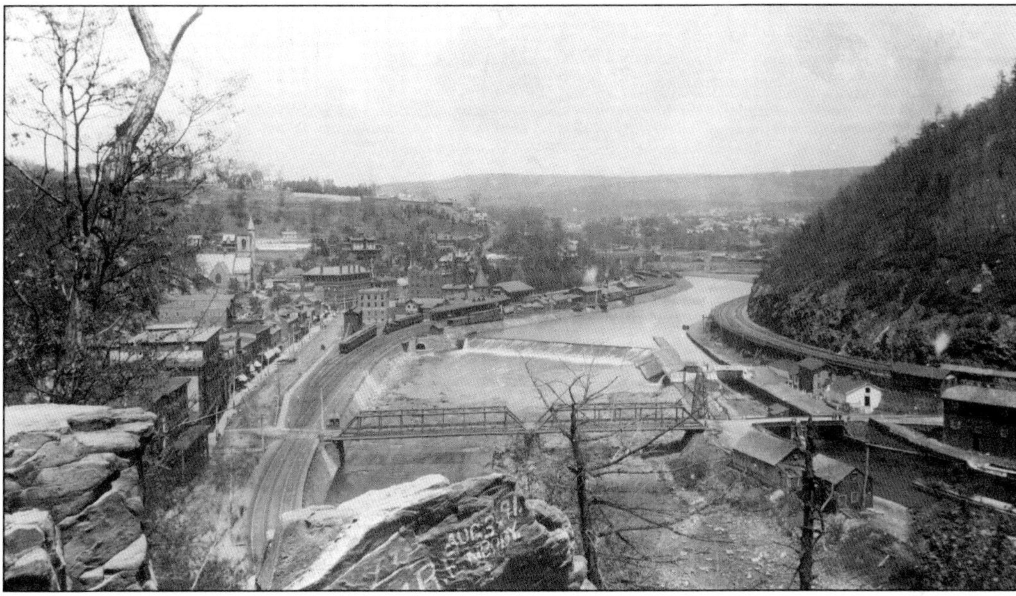

The 12.5-foot-high Mauch Chunk Dam (center) formed the south end of Mauch Chunk harbor. The lower division of the Lehigh Canal entered the harbor above the dam (center right). A footbridge, destroyed in the 1901–1902 floods, enabled Lehigh Valley Railroad passengers to cross the river from the train station at the foot of Bear Mountain (right) to Mauch Chunk. (Courtesy Beverly Sillers.)

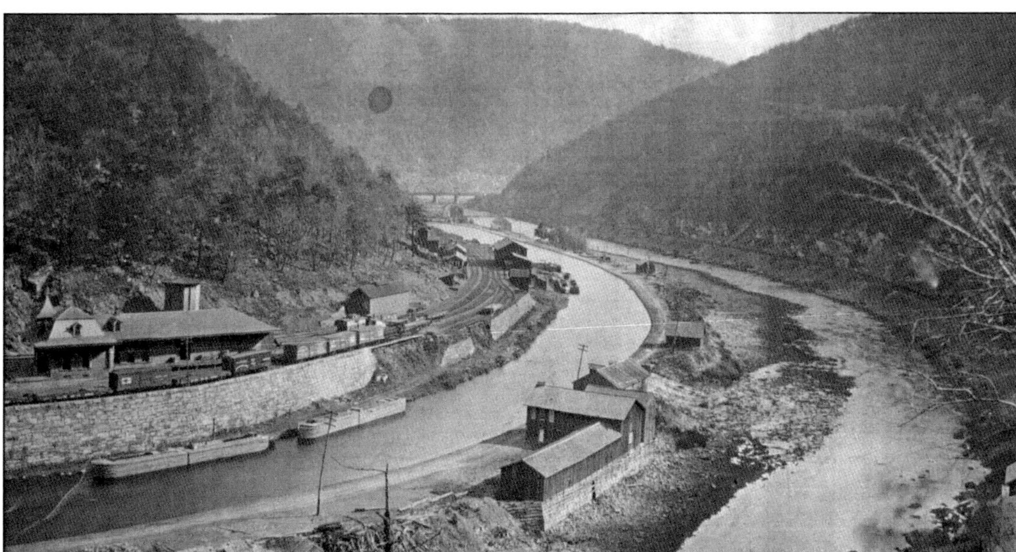

The Lehigh Valley Railroad's first station lies behind a stone embankment at the foot of Bear Mountain in this scene looking down the Narrows. The railroad's tracks paralleled the canal on the east (left) side of the river and crossed to the west side by bridge. The tracks of the Central Railroad of New Jersey ran along the west bank of the Lehigh beneath the steep cliffs of Flagstaff Mountain.

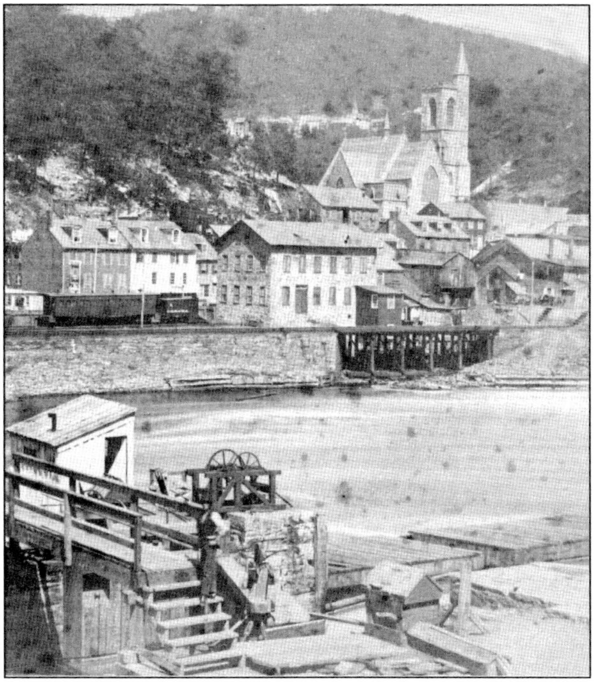

The Lehigh Canal entered Mauch Chunk harbor through Guard Lock 1 (foreground). Depending on the water level in the Lehigh, the average lift of the lock was 1.4 feet. The Mauch Chunk Creek, which passes under Mauch Chunk through a culvert, entered the Lehigh River beneath the trestle. The Hooven Building is to the left of the trestle, and St. Mark's Episcopal Church is in the background. (Courtesy Ray Holland.)

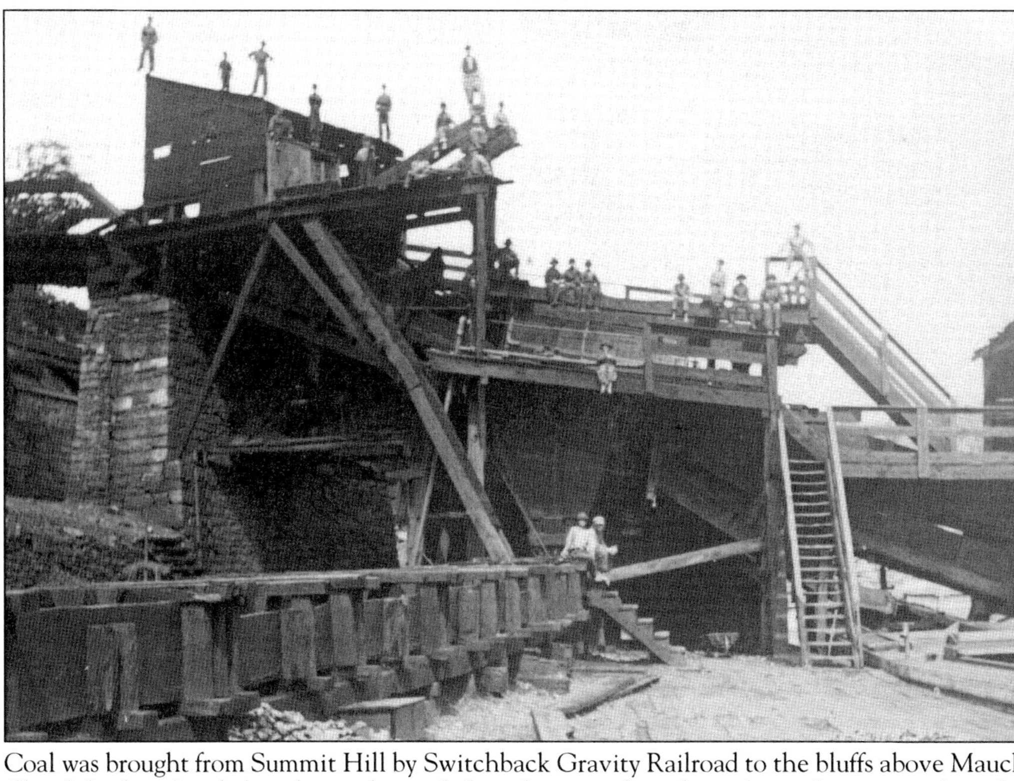

Coal was brought from Summit Hill by Switchback Gravity Railroad to the bluffs above Mauch Chunk harbor. Loaded coal cars descended to the waterfront by a short plane, providing the power to pull the empties back up the incline. The coal was tipped into a reservoir and loaded onto boats by a chute. (Courtesy Ray Holland.)

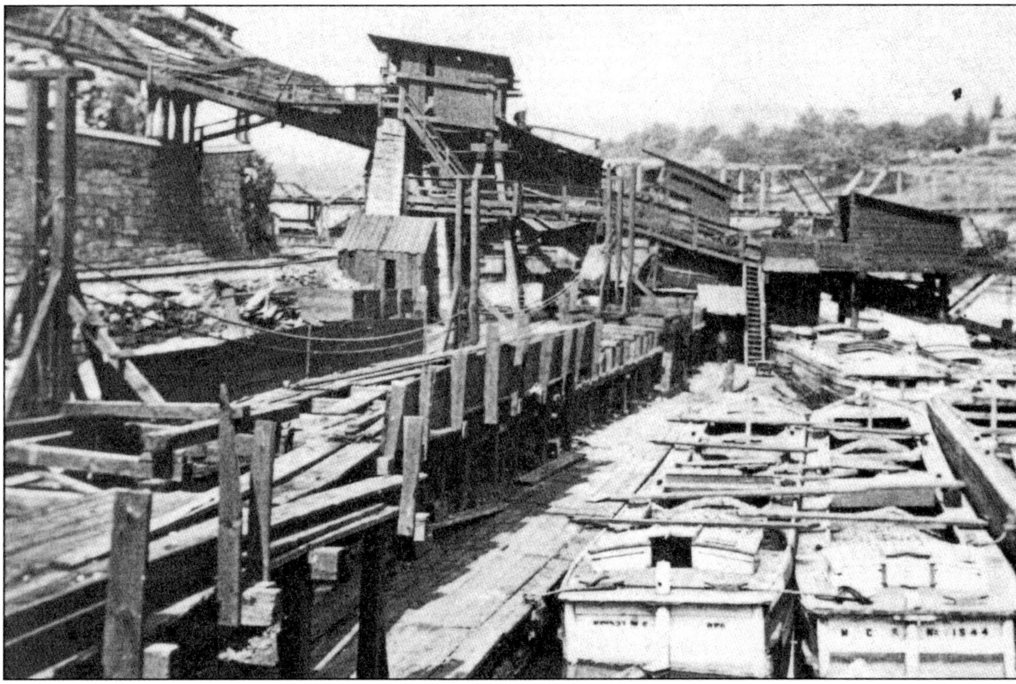

Coal boats wait to be loaded at the foot of the Mount Pisgah coal chutes.

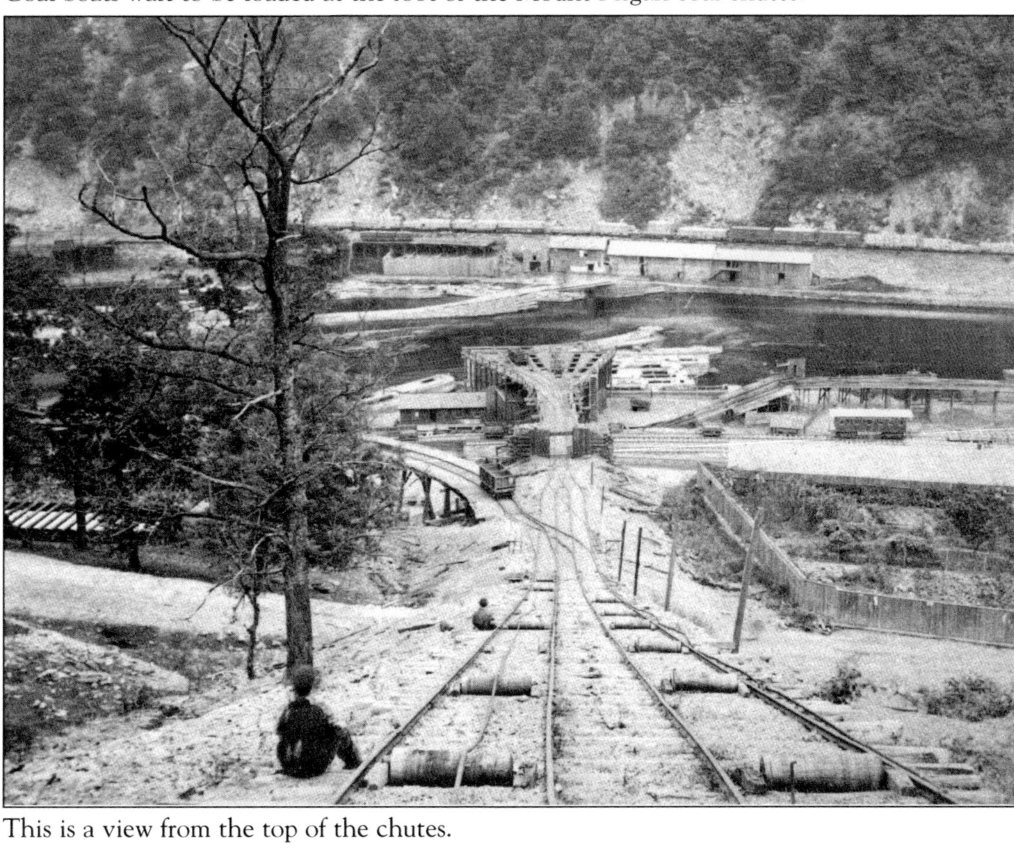

This is a view from the top of the chutes.

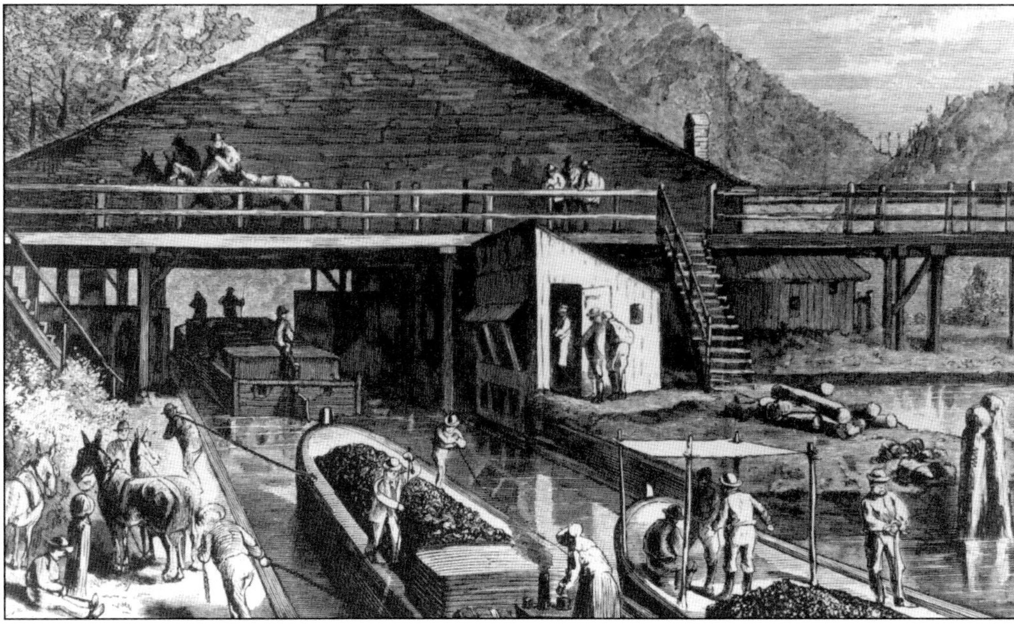

Lock 2, Lower Division, was a side-by-side standard lock and weigh lock. The dual construction kept empty boats moving up the canal while loaded boats, descending the canal, were weighed. To weigh a coal boat, the weigh lock was emptied and the boat lowered onto the scale. The maximum load was 100 tons. This pen-and-ink drawing dates from 1870.

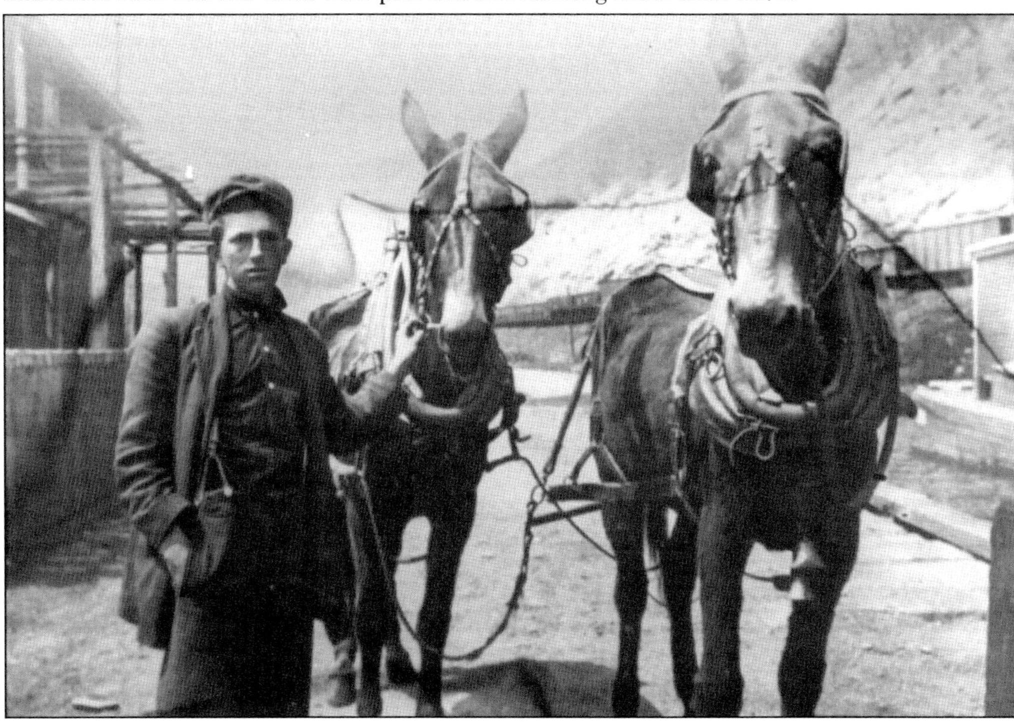

Canal boats were drawn by mules. Each captain owned his own team of mules. During the off-season, the mules were boarded with a farmer who fed and cared for them in return for work around his farm.

A coal boat passes through a lock at Lehigh Gap *c.* 1920. With the exception of Lock 1 near Mauch Chunk (two locks in tandem with a four-boat capacity), locks on the Lehigh Canal measured 100 by 22 feet, wide enough to hold two coal boats traveling in the same direction. They were filled at the upper end through Josiah White's invention, the drop gate, and emptied at the lower end through conventional mitre gates.

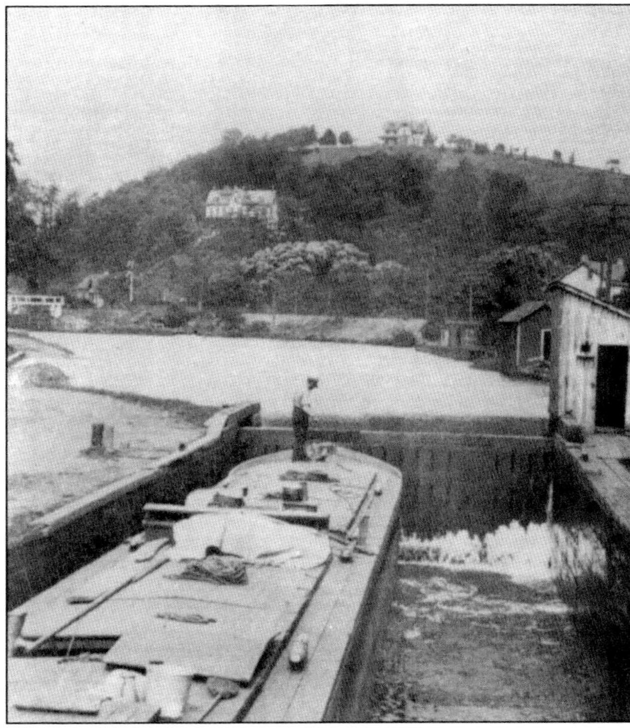

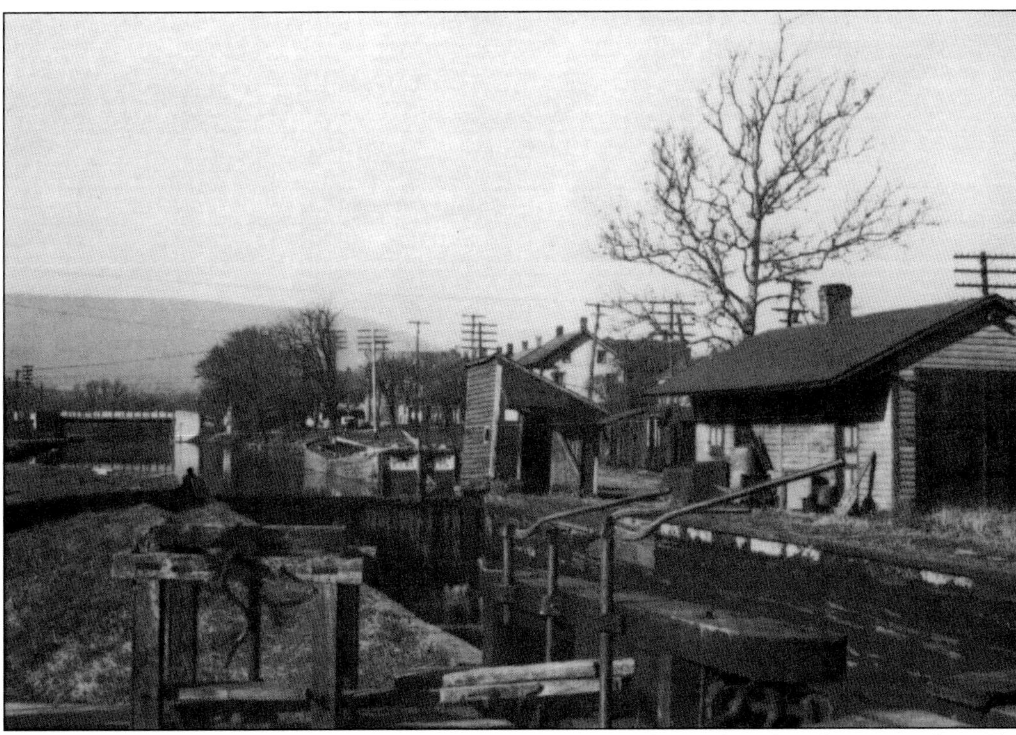

A lock was known to boatmen by the name of the lock tender, rather than by its number. This is Kelchner's Lock in Walnutport. Mitre gates are in the foreground; the drop gate is at the upper end of the lock.

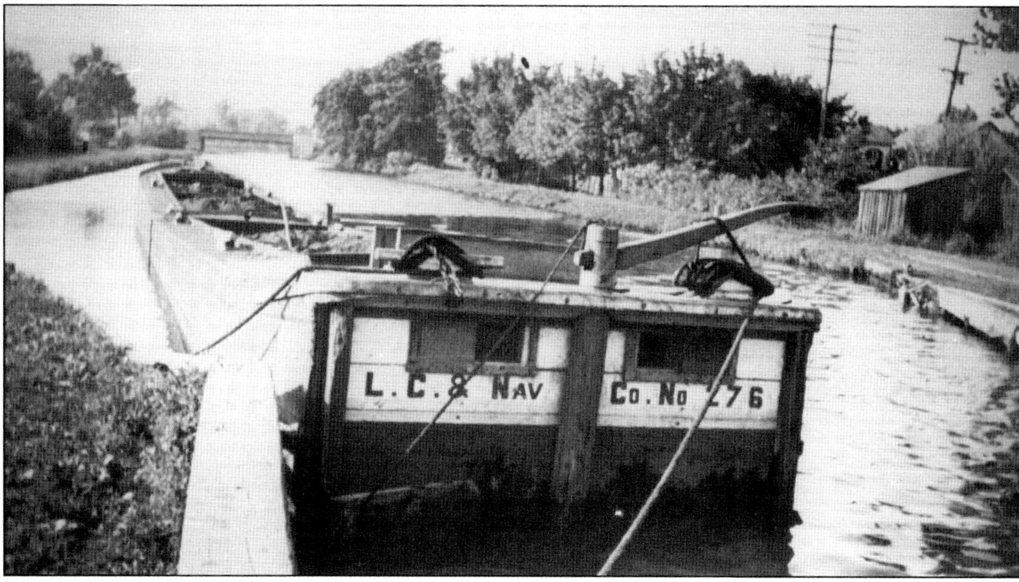

Drawn by a two-mule team and controlled by a tiller and rudder, coal boats were operated by either a two-man crew of captain and mule driver or by a family. A wife relieved her husband at the helm, and she and her children, as young as 10 years old, drove the mules along the towpath. A Lehigh Coal & Navigation coal boat was 87 1/2 feet long and 10 1/2 feet wide. A 10-by-10-foot cabin was located in the stern. Early coal boats (stiff boats) were one continuous length. At the end of the 19th century, coal boats were sectioned into two hinged boats to facilitate turning, loading, and off-loading.

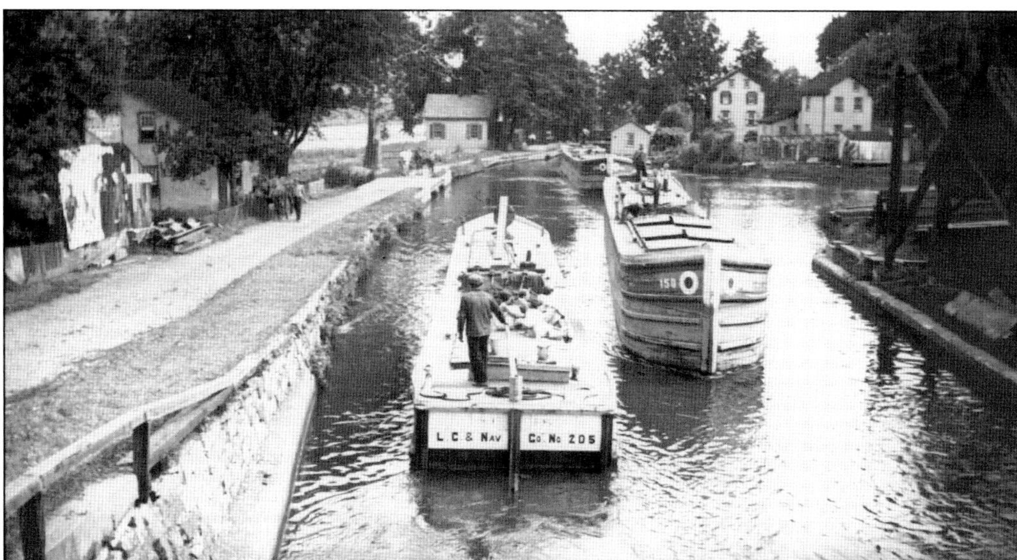

An empty boat passes a loaded boat. The payload was 100 tons. When loaded, a coal boat weighed approximately 120 tons and had a five-foot, three-inch draft.

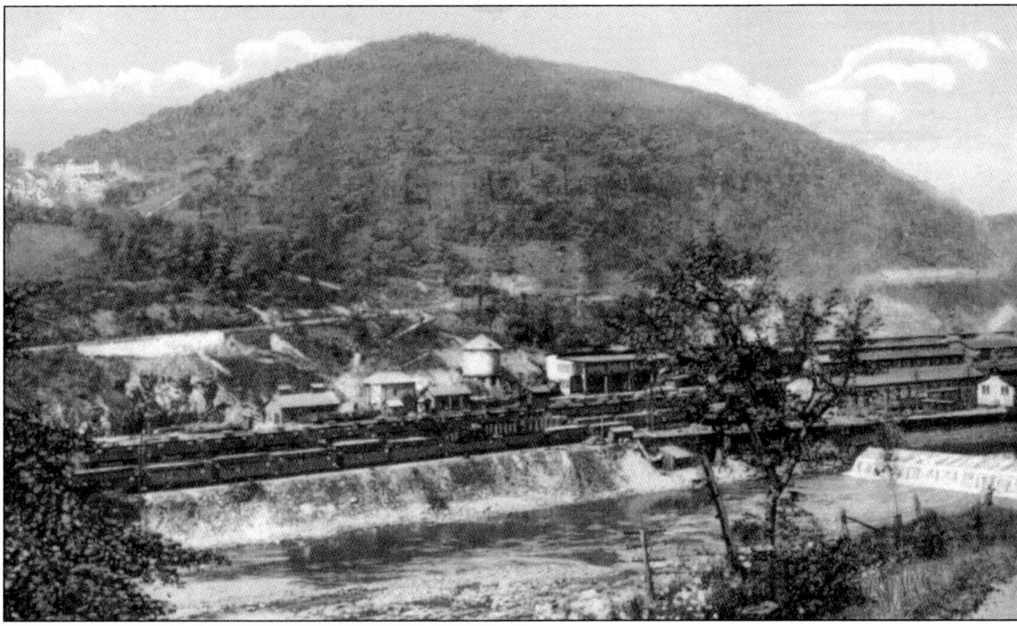

An even greater feat of engineering than the Lower Division was the construction of the Upper Division of the Lehigh Canal in 1835. It ran north to White Haven through a 26-mile wilderness of mountain, rock, and wild water. Twenty enormous dams and 29 locks with extremely high lifts overcame the 600-foot drop in elevation between White Haven and Mauch Chunk. The 200-foot-long, 20-foot-high Packer Dam crossed the river at the entrance to the Upper Division. Its lock had a lift of 15.33 feet.

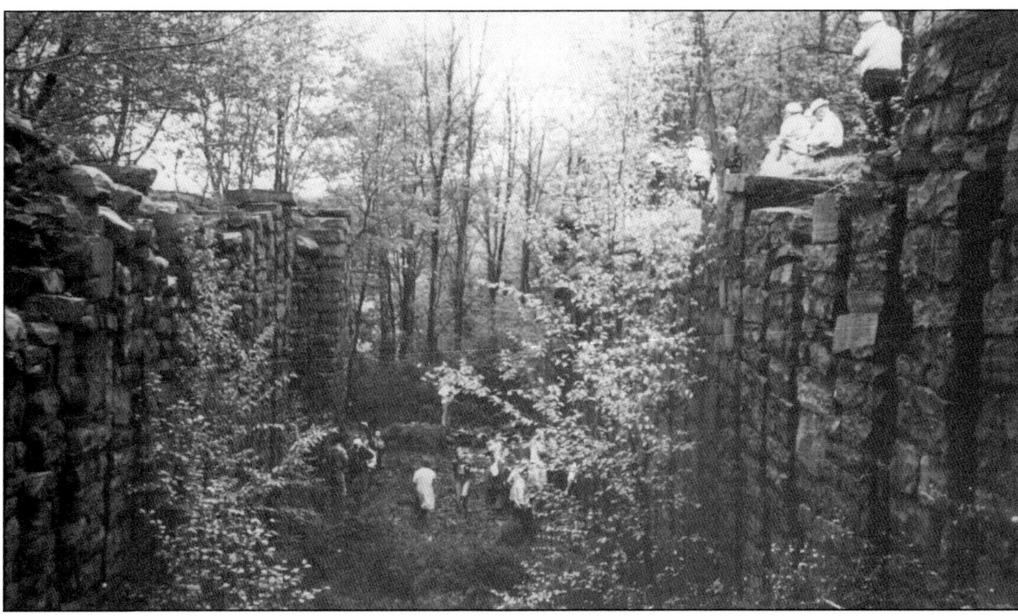

Tourists explore the remains of the 30-foot-high Lock 27 at Lehigh Tannery. With the steep drop in elevation, the Upper Division was extremely vulnerable to flooding. Rebuilt after the 1841 flood, it was destroyed beyond repair in 1862 and replaced by the iron horse, running along an extension of the Lehigh Coal & Navigation Company's Lehigh & Susquehanna Railroad.

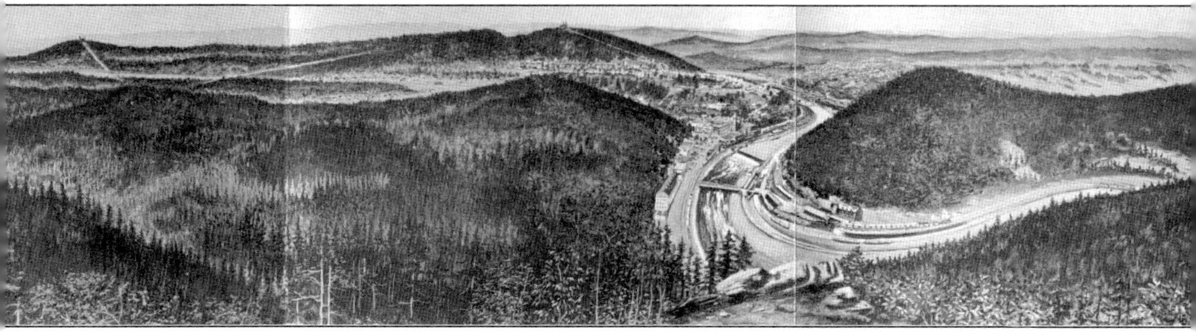

As coal was not moving fast enough along the single-track gravity railroad, a back track was constructed in 1845. Cars ascended Mount Pisgah by an inclined plane (right), crossed a trestle, and descended by gravity down the other side of Mount Pisgah to the base of Mount Jefferson. Another inclined plane took the cars up Mount Jefferson (far left), and they descended by gravity to Summit Hill. The return trip to Mauch Chunk was by gravity along the original single-track railroad. Engine houses, containing machinery to operate the planes, are visible on the summits of the two mountains.

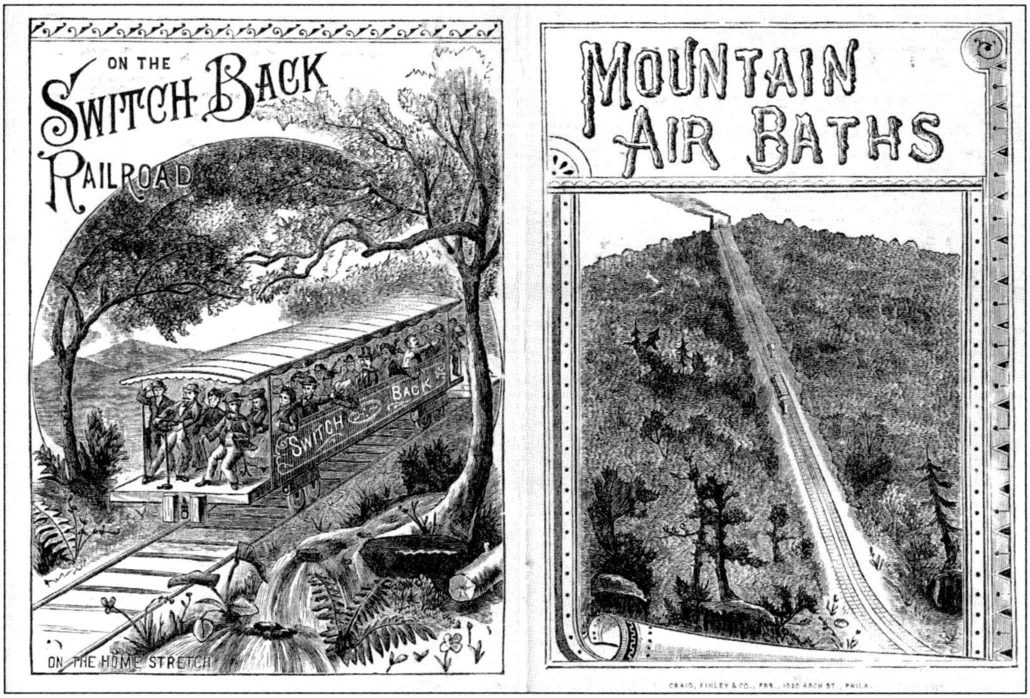

Late-19th-century advertising popularized the Switchback as it became exclusively a tourist thrill ride in 1872.

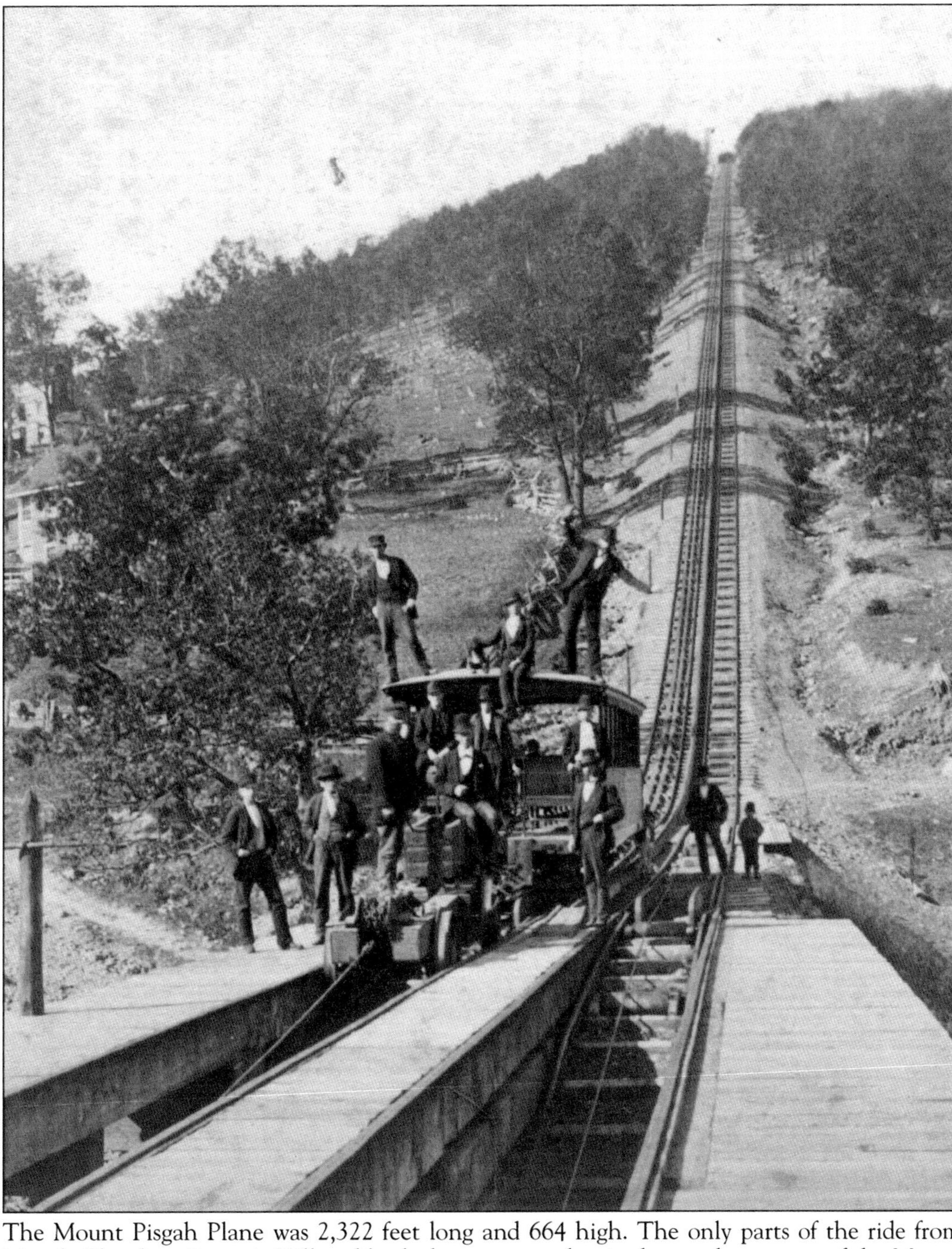

The Mount Pisgah Plane was 2,322 feet long and 664 high. The only parts of the ride from Mauch Chunk to Summit Hill and back that were mechanized were the ascents of the Mount Pisgah and Mount Jefferson Planes. The rest was a freewheeling thrill by gravity, with a single brakeman in control. (Courtesy Ray Holland.)

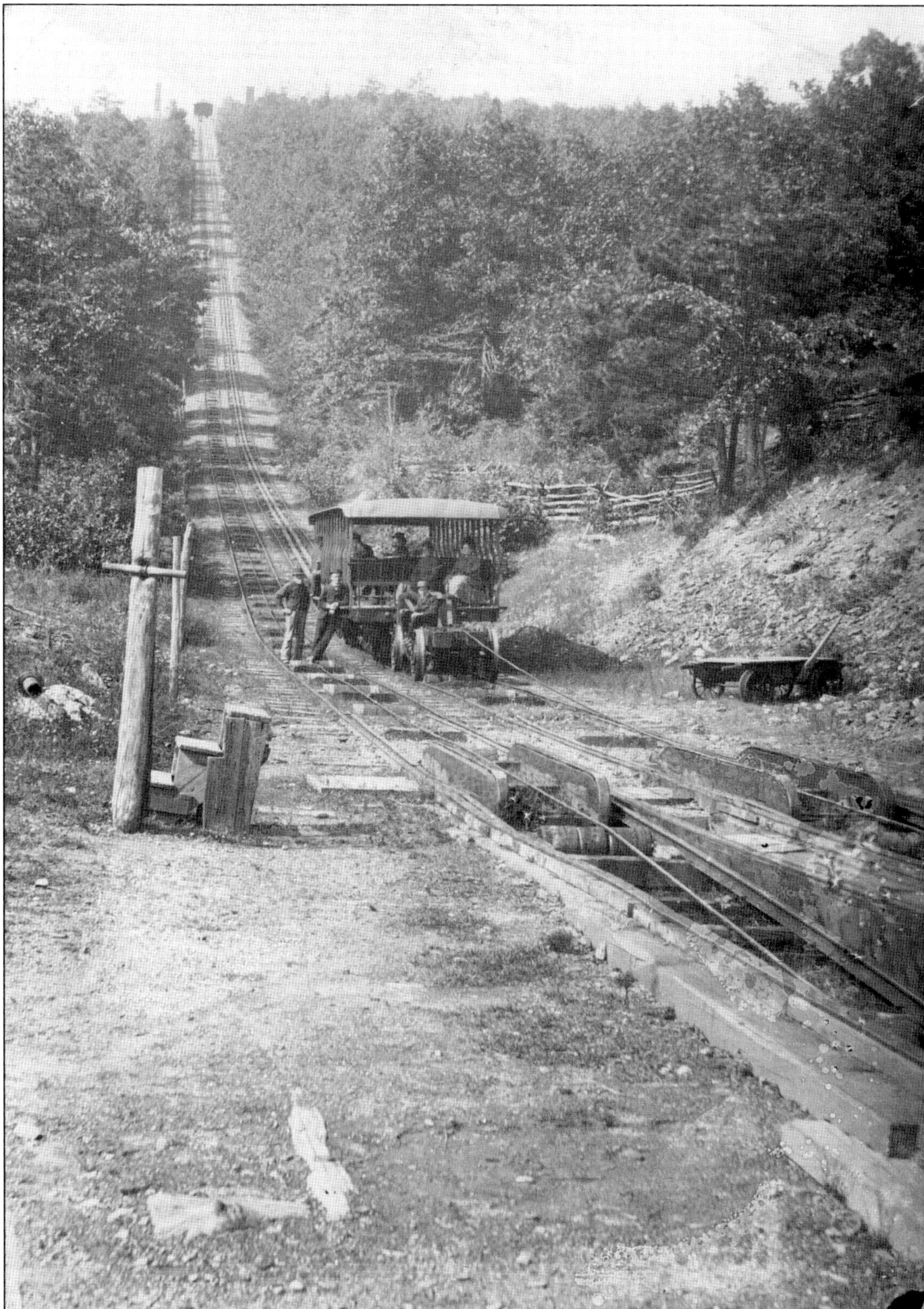

The Mount Jefferson Plane was 2,070 feet long and 462 feet high. (Courtesy Beverly Sillers.)

A booster (called a safety car by Josiah White but known colloquially as a barney car) was housed in a pit at the foot of each plane. Hauled up the plane by iron bands (driven by a giant wheel in the engine house at the head of the plane), it pushed passenger or coal cars up the incline. (Courtesy Ray Holland.)

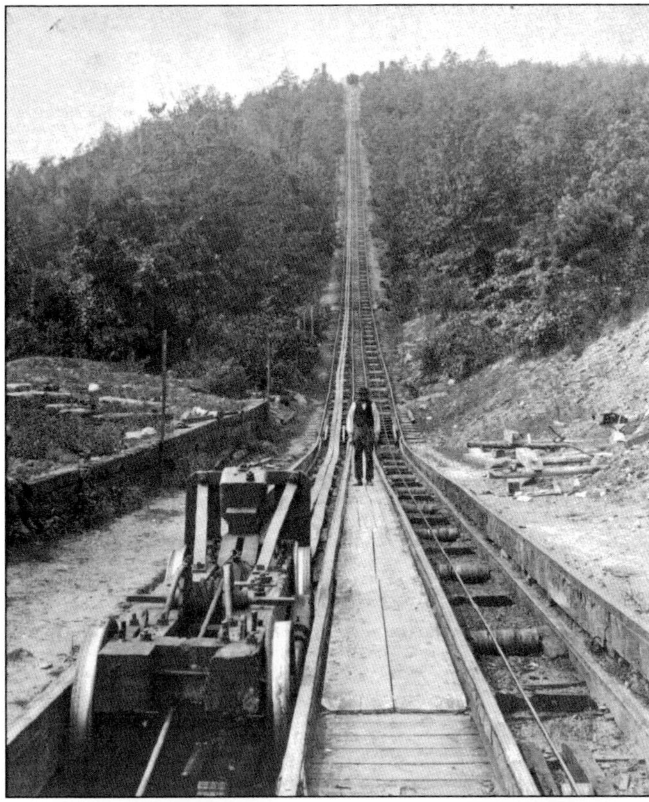

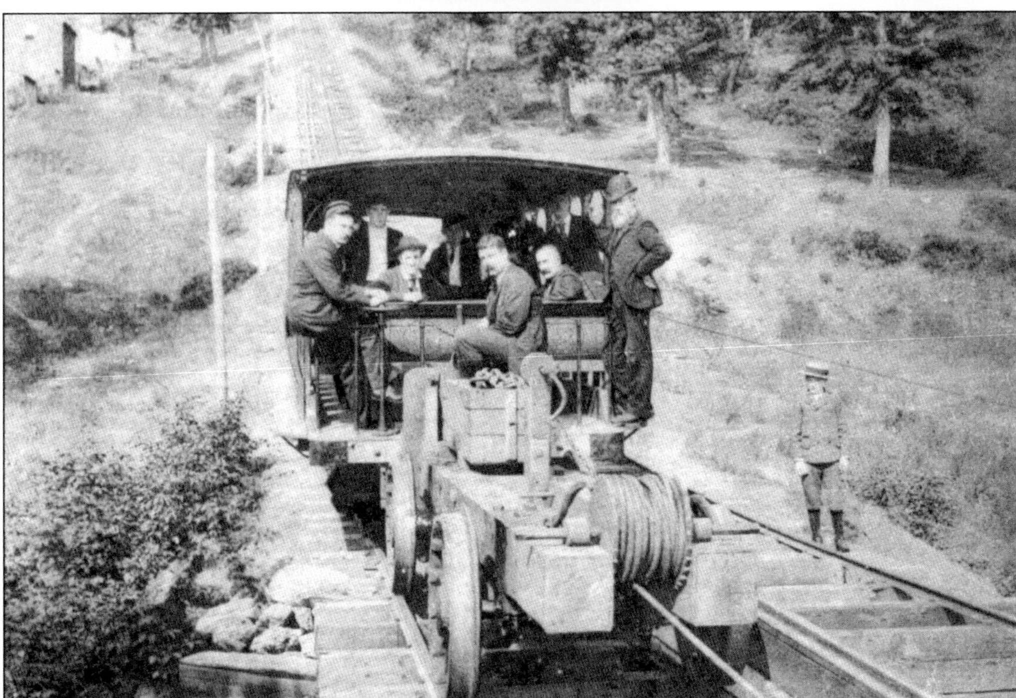

The barney car moves into place behind a passenger car.

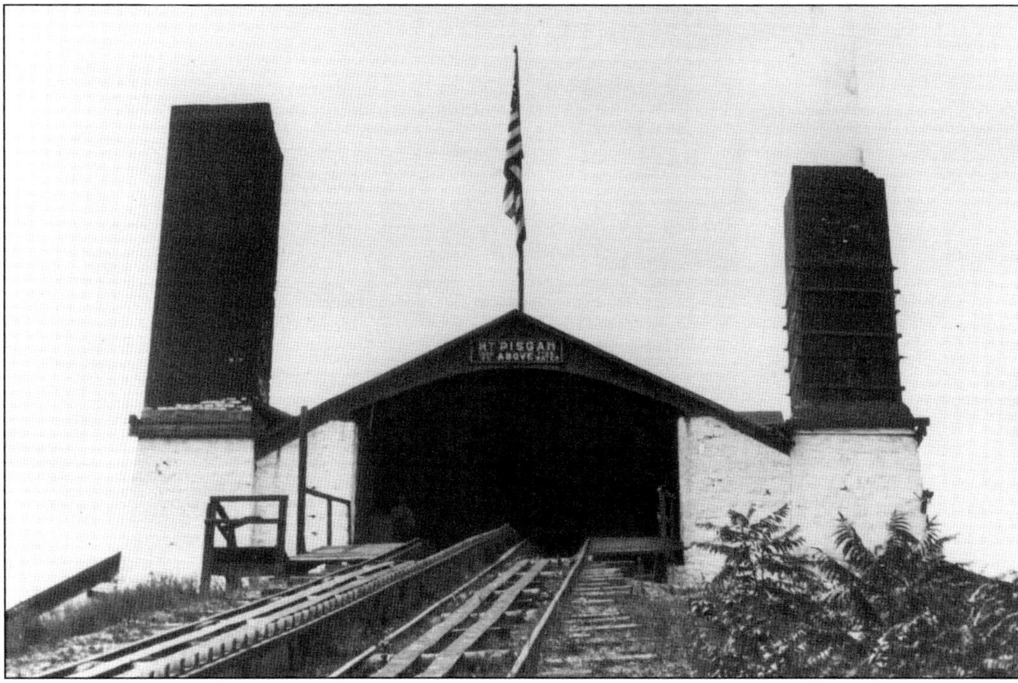

A stationary steam engine, located in an engine house on top of each plane, provided motive power for the barney car. This is the Mount Pisgah Engine House.

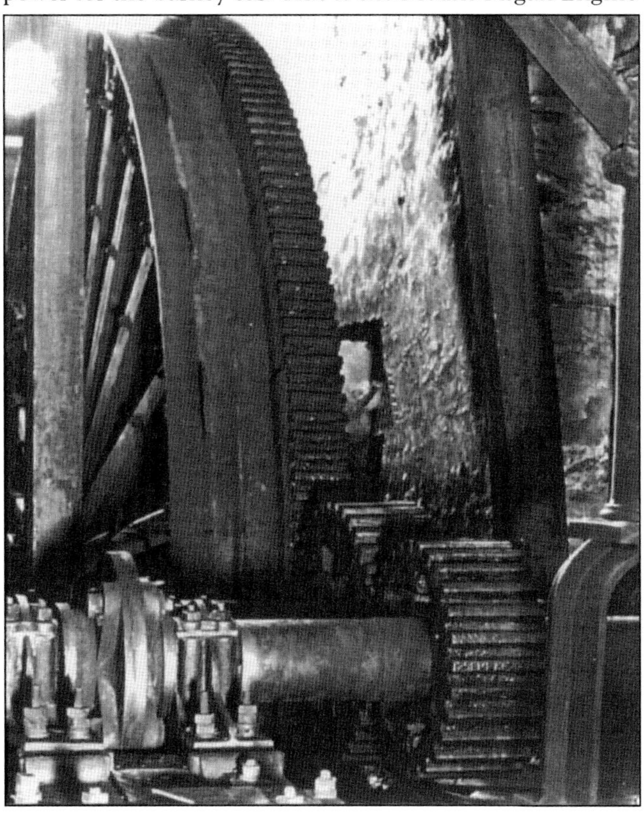

A steam engine drove a 28-foot drum, which reeled in two Swedish iron bands, each $7^1/_2$ inches wide, to which the booster was attached.

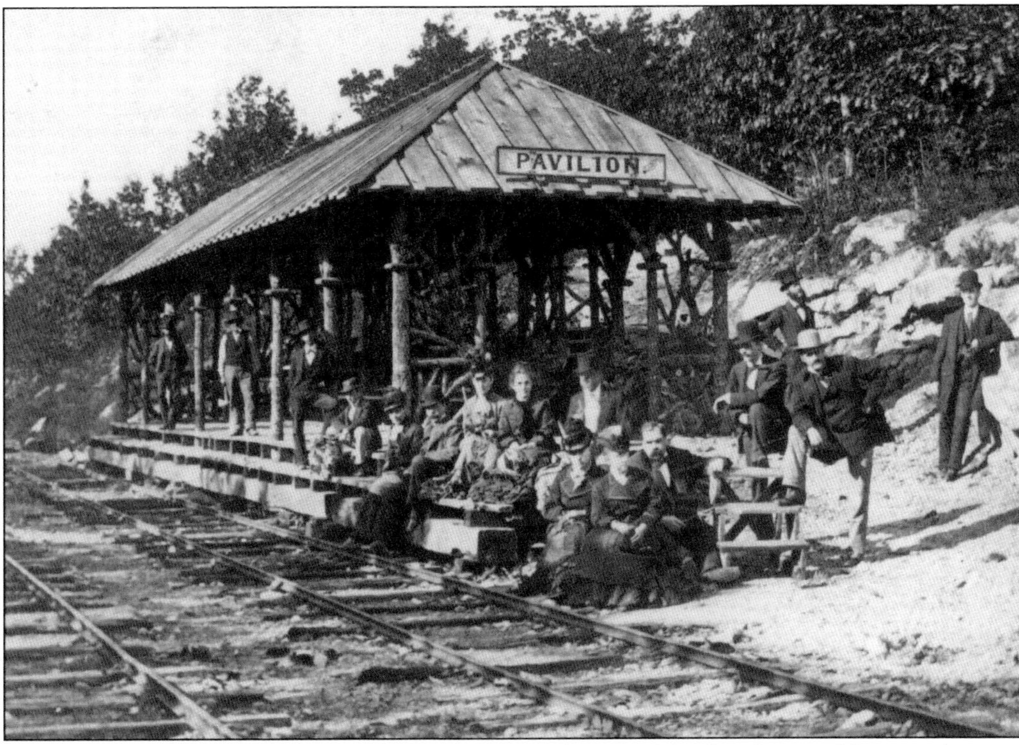

At the Pavilion Station at the top of Mount Pisgah, Switchback riders could enjoy the view and eat lunch. (Courtesy Ray Holland.)

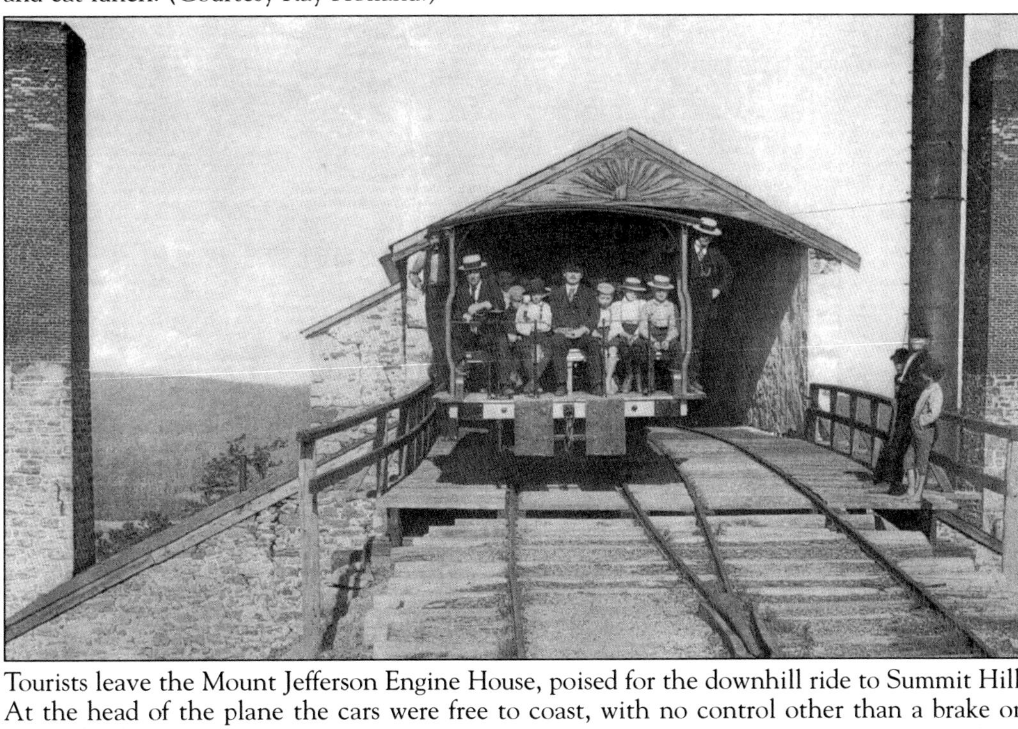
Tourists leave the Mount Jefferson Engine House, poised for the downhill ride to Summit Hill. At the head of the plane the cars were free to coast, with no control other than a brake on their wheels.

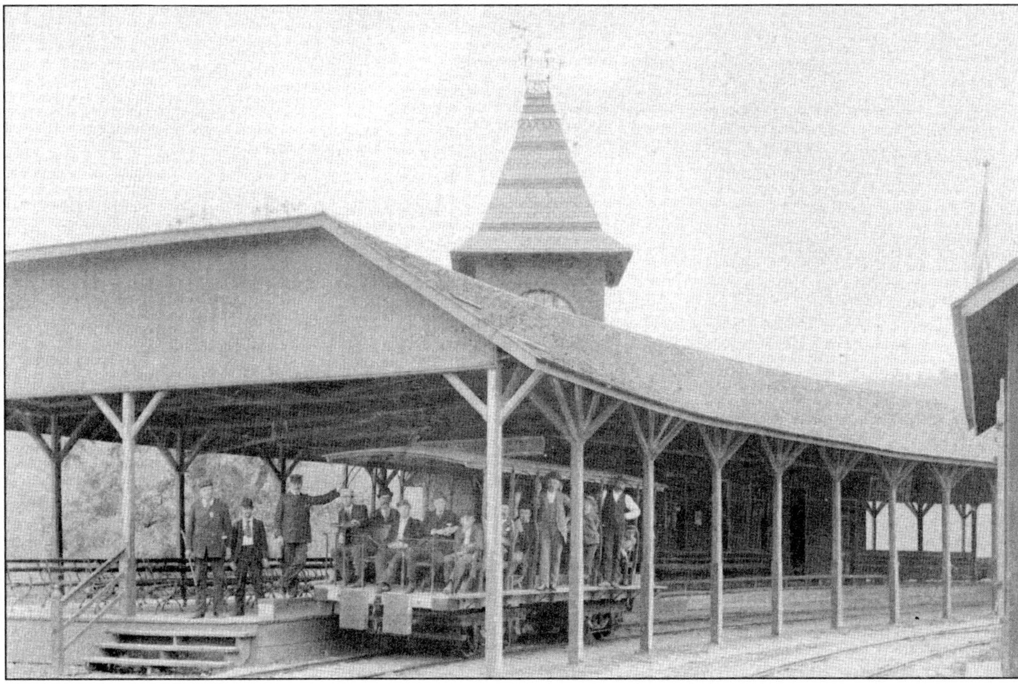

Tourist Walter C. Quevedo rode the Switchback from Mauch Chunk to Summit Hill and back in the late 19th century. He recounts each stage of the trip in his diary. Embarking at Mauch Chunk Station, he writes, "We boarded a car and the brakes were given a holiday."

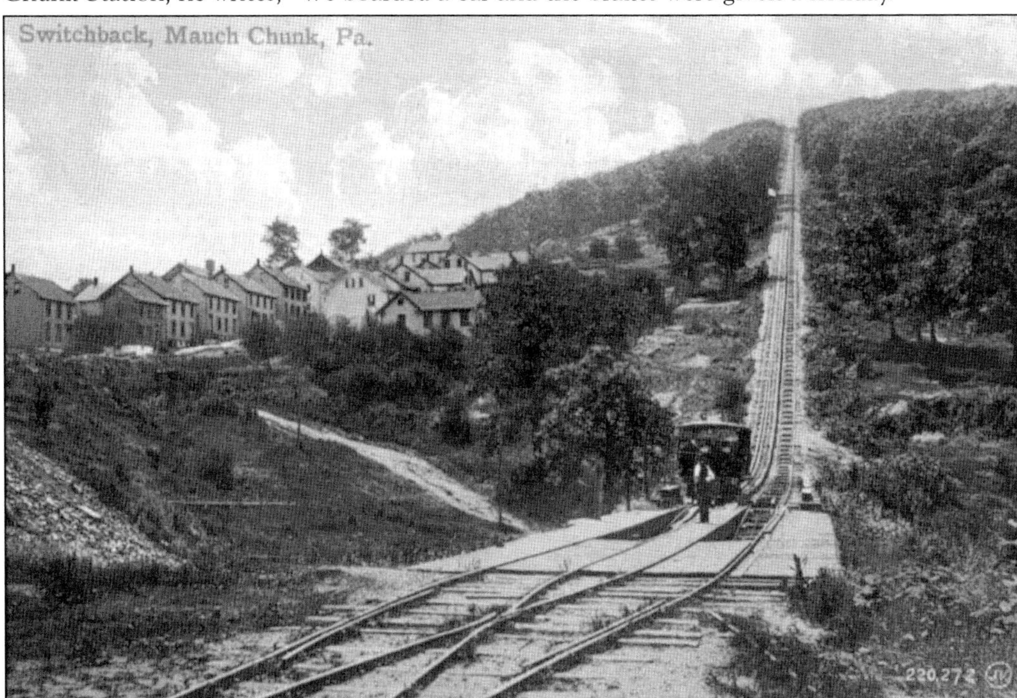

Quevedo continues, "The car proceeded on its own until it came to the bottom of the mountain—the foot of Mount Pisgah plane."

The passenger car crosses the barney pit. The barney car is drawn from the pit to push the passenger car up the plane. Quevedo writes, "At this point a flat car, filled with pig lead and other animal matter, slid under us and began to boost us to the top, a distance of 2322 feet."

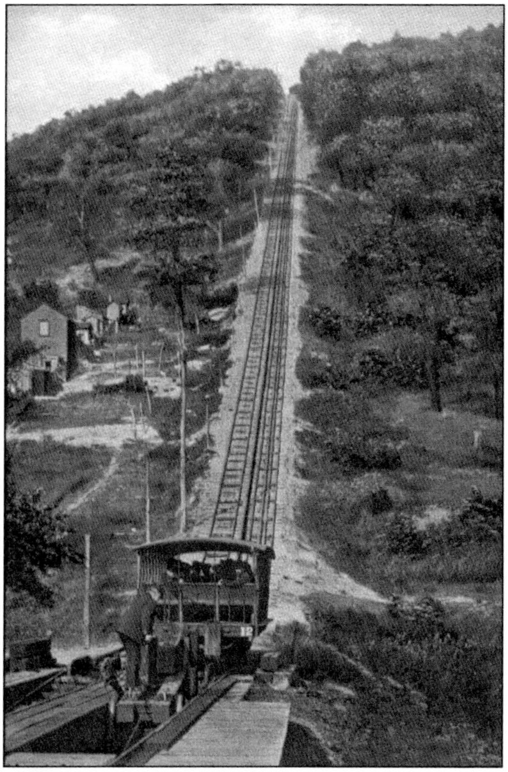

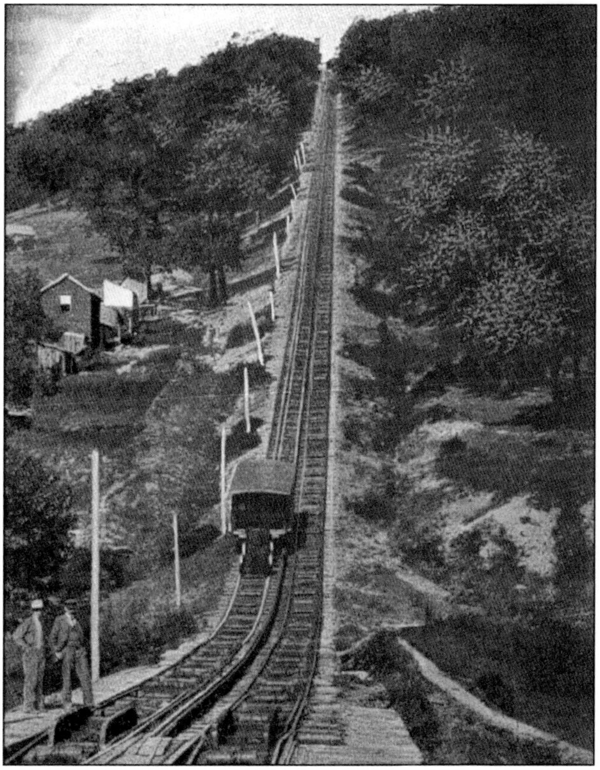

As the Switchback car ascends, he observes, "the incline is so steep, we almost stood on our hair while doing the angel act. If our readers will place their feet on the mantel-piece and imagine that the house is climbing over the house next door, they will have a slight idea of the position we assumed."

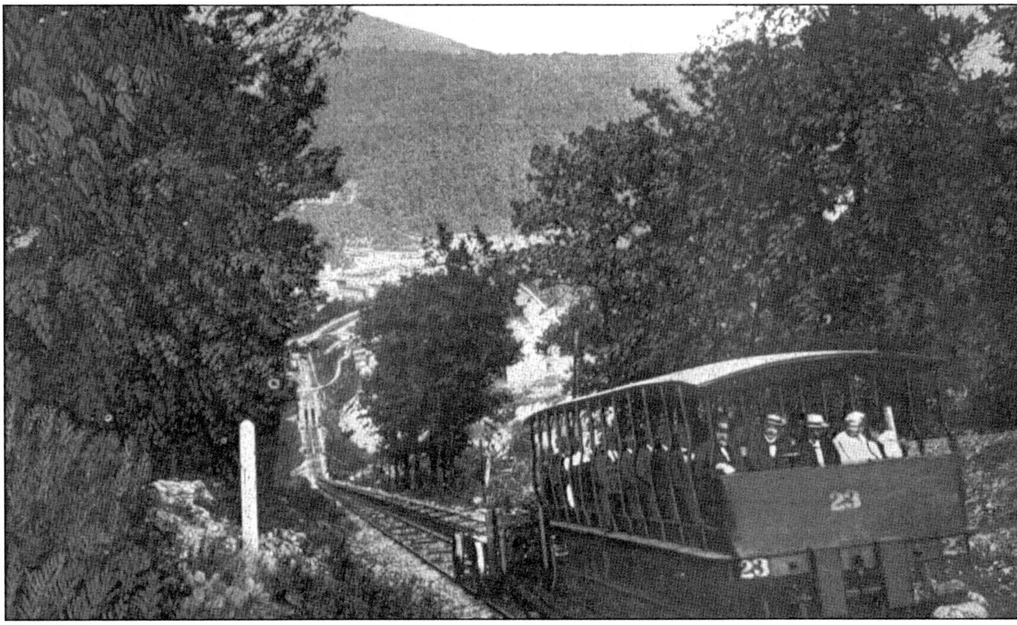

Approaching the summit of Mount Pisgah, Quevedo continues: "The booster was drawn by a cable, and if its suspenders had ever bursted, we would have had never more use for false teeth and ear muffs. Our hair raised our hat, until it looked as if it was stuck up there for fun, as the conductor stated, when we arrived at the roof of Mount Pisgah, that we were 1,500 feet above tidewater."

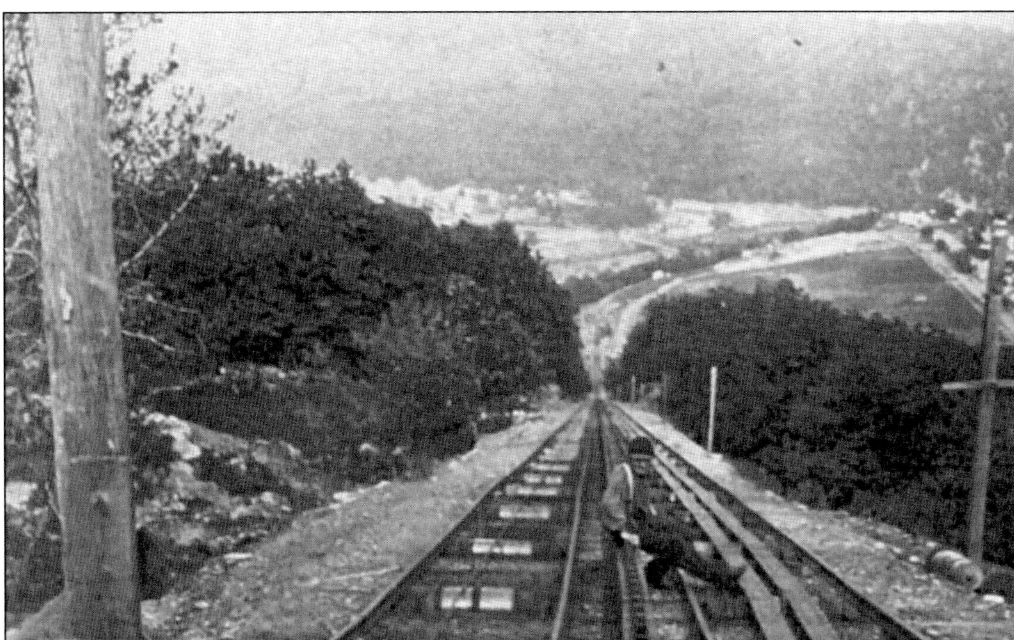

Quevedo describes the view toward Mauch Chunk from the top of Mount Pisgah: "The scene from Mount Pisgah is excelled by none in the country. There is perhaps, not another place in the world where so much of the truly sublime and picturesque is so magnificently displayed as here."

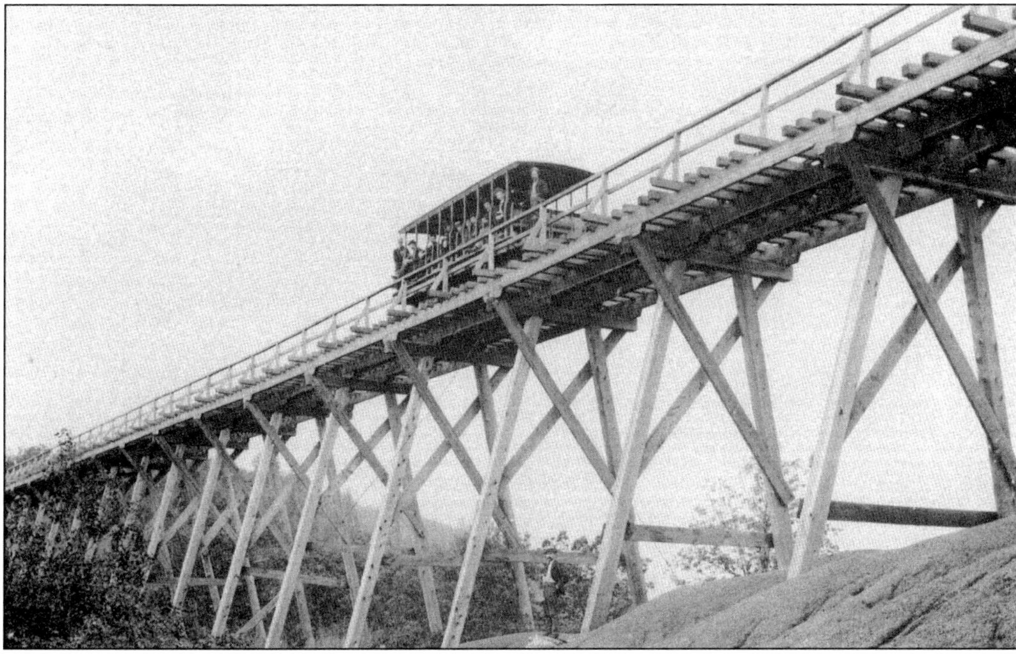

Quevedo continues: "We passed through the engine house and slowly over a trestle spanning a wild ravine. To the right and north is Broad Mountain and, in a ravine, halfway between the summit and the river, is Glen Onoko, a piece of Nature's handiwork."

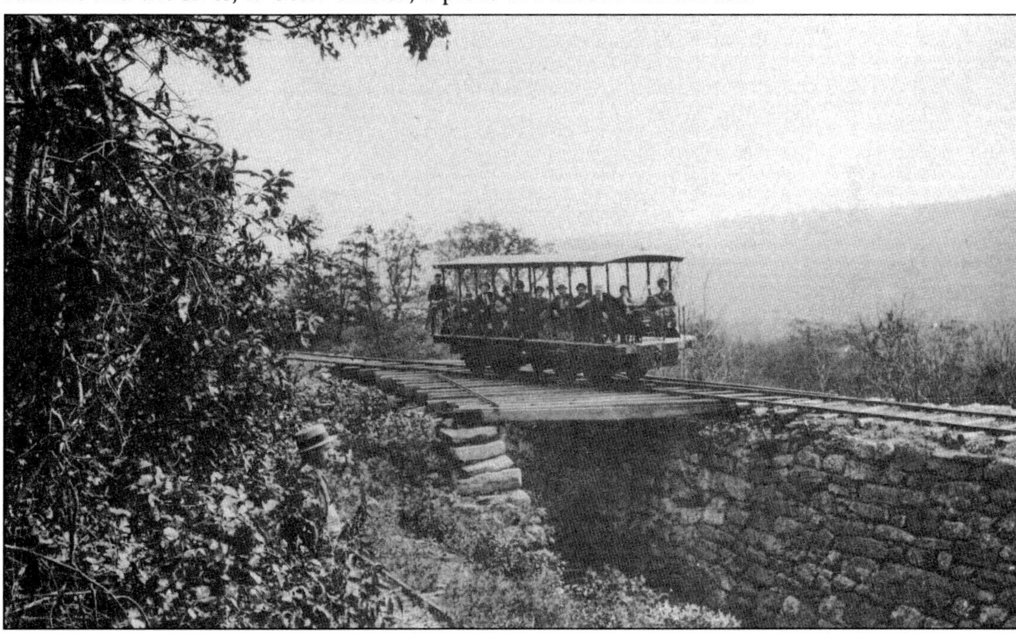

As the train crosses over the Switchback down the tracks at Five-Mile Tree Bridge, Quevedo writes: "After we feasted our eyes on this scene, the young man who officiated at the brake, got tired and let go. Then we scooted down the offside of the mountain at the rate of about a million miles an instant. We thought of all the sins we had ever committed, held on to our seat with one hand, our hat with the other, and wished that we'd minded our own business and stayed at home."

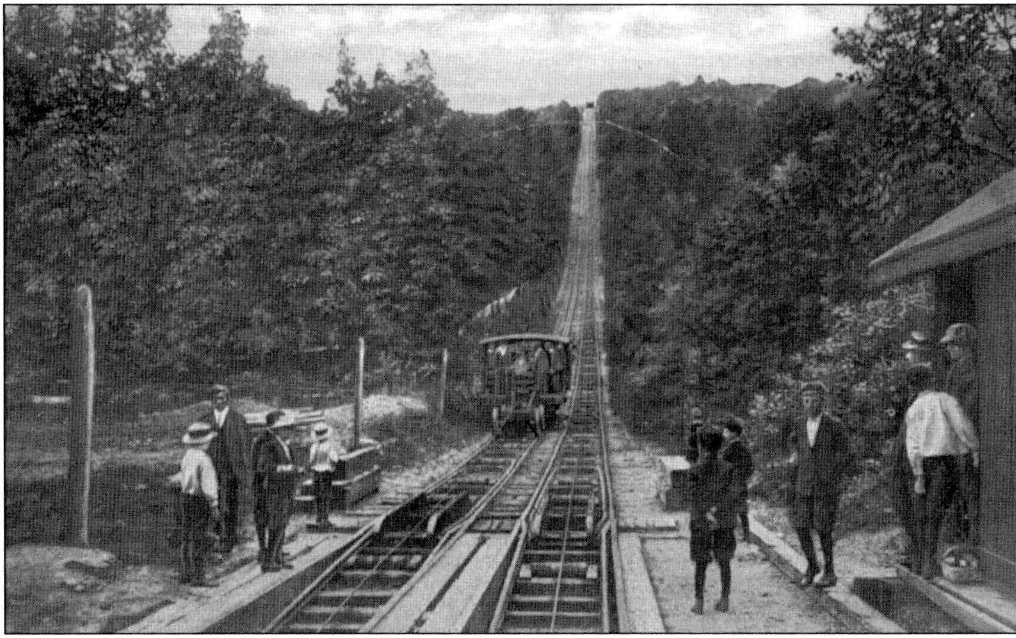

Moving across the barney pit to the base of the Mount Jefferson Plane is described by Quevedo: "The booster again came to our rear and we were shoved up to the top, with our heels in the air, and our head under the seat."

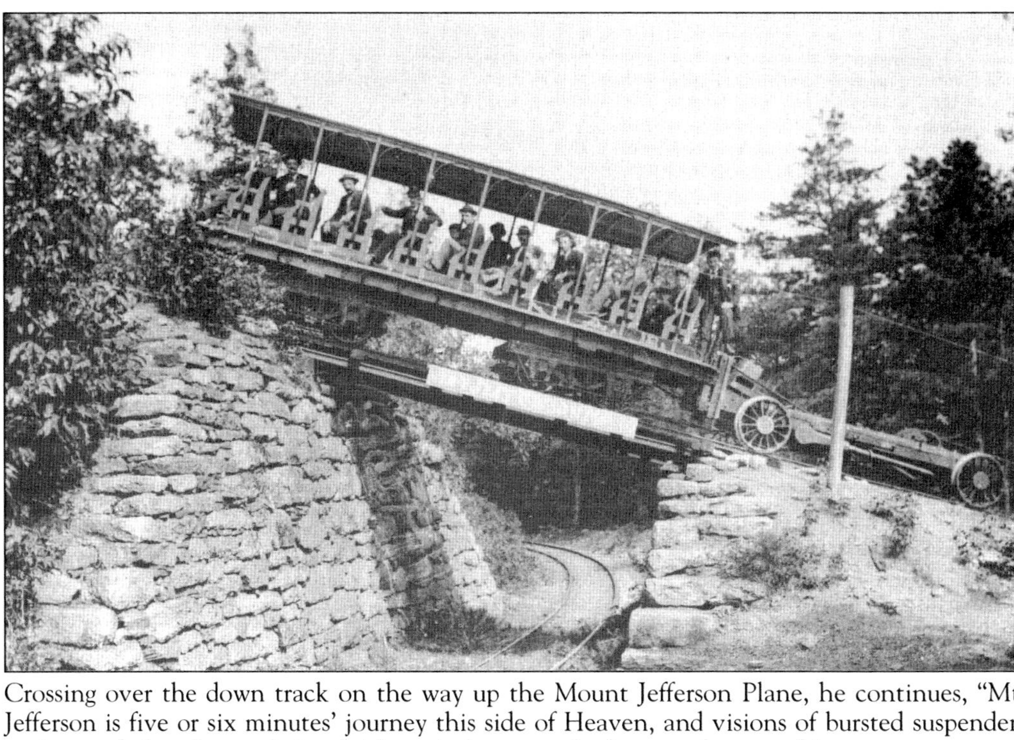

Crossing over the down track on the way up the Mount Jefferson Plane, he continues, "Mt. Jefferson is five or six minutes' journey this side of Heaven, and visions of bursted suspenders again crawled under our hair, which again began to do duty as a hat stand."

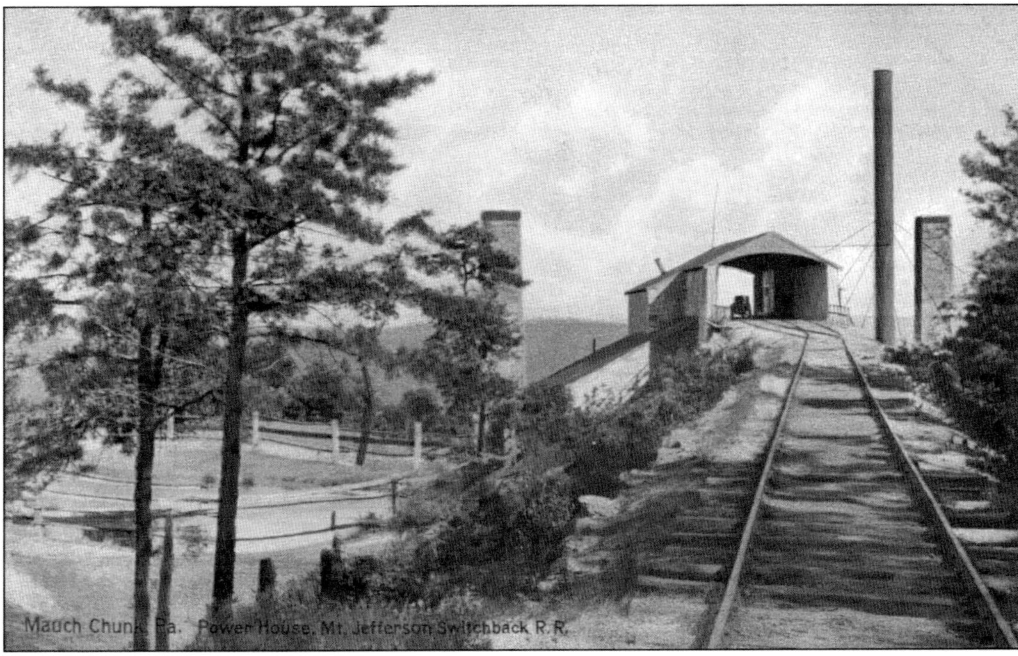

On reaching the summit of Mount Jefferson, Quevedo notes, "We were now 1660 feet above tidewater."

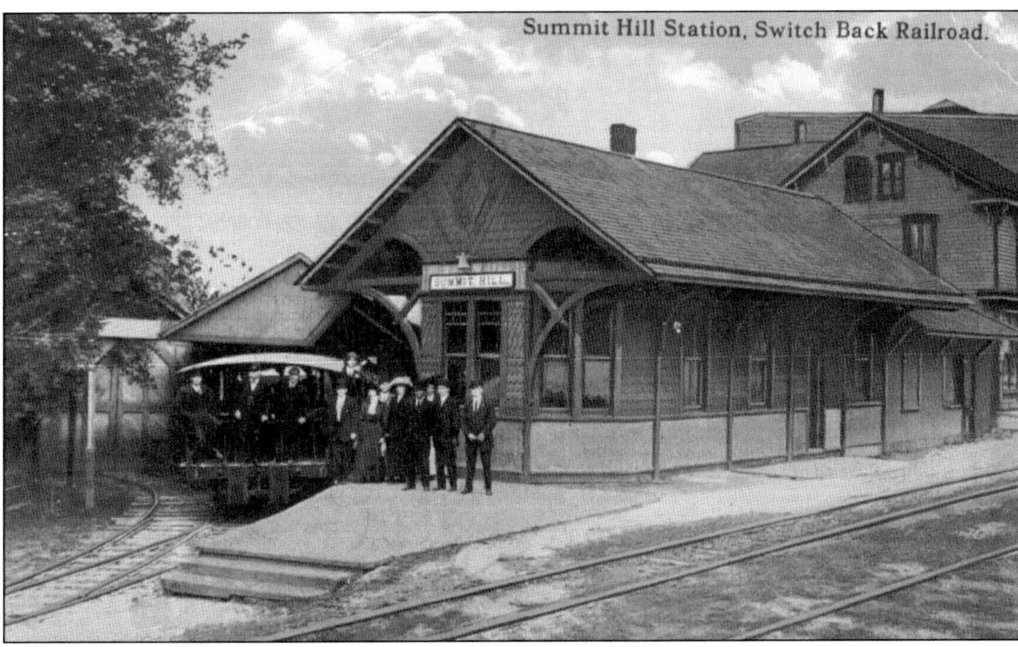

Recalling the ride down Mount Jefferson to the Switchback Depot on Summit Hill, he writes: "After the young man, who had no fear of death, had spat on his hands, he gave the brake another vacation and we began to saunter down the mountainside like—our readers must pardon us—we came very near using unorthodox language!"

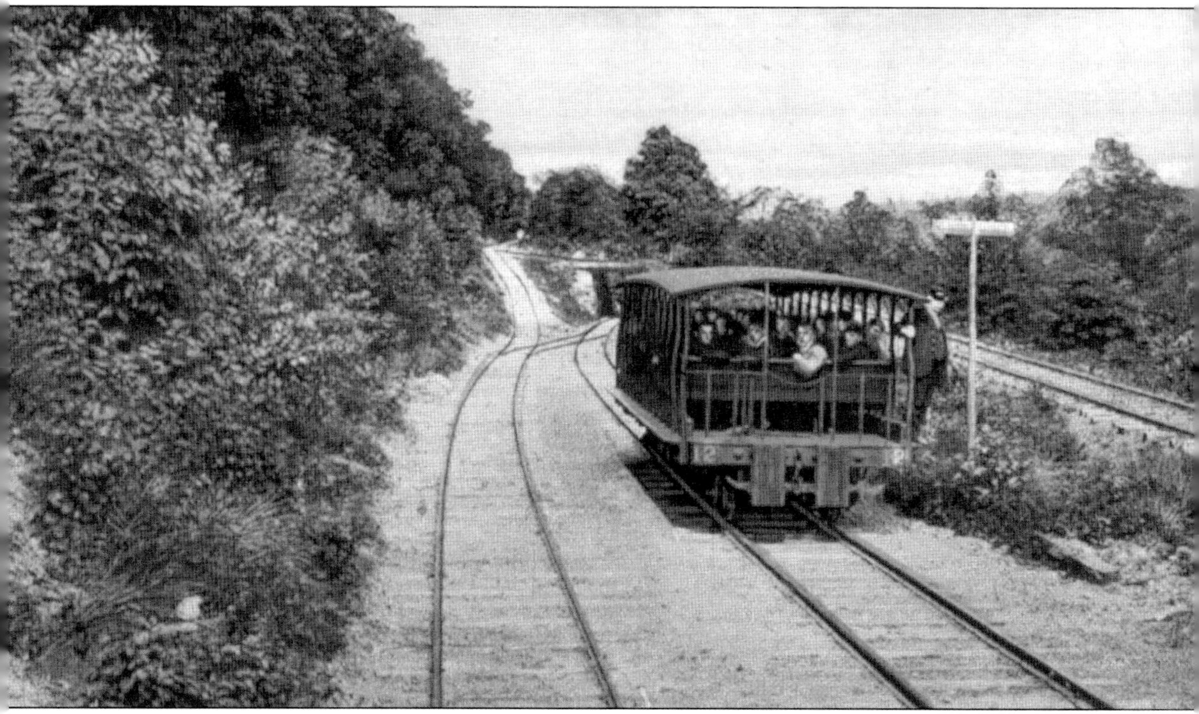

Riding back to Mauch Chunk along the downhill track, when speeds often reached 65 miles per hour, Quevedo concludes: "The brakes were again excused, and down we started on the return journey making a detour of the two mountains. If we had made a swift journey to Summit Hill, it was nothing but a crawl compared with the homeward trip. And while we were riding on this hell-bent trip, the people in the thriving town below didn't appear to mind it a bit, and didn't appear to appreciate that, at any moment, we might come tumbling down upon them and knock them into smithereens."

Down the Switch Back

Side by side we rode together,
On a clear October day,
While the mountains, crimson-crested,
Kept a royal holiday.
Down the Switch Back from Mount Pisgah
We went speeding o'er hills,
With the golden sunlight flashing
From the rippling mountain rills.
But the flashing and the glinting
And the blue of autumn skies
Were but frosty in their beauty
To the summer of her eyes.
Side by side, we rode together,
And I did not dare to wait,
For she was seventeen, and I
Was turned of forty-eight

So I whispered to her: "Darling,
Let us travel, side by side,
Down the grade of Life's long Switch back,
To the shoreless ocean's tide."
But she looked away, far over
All the hills that lay between,
To the distant, dim horizon
And her eyes were *too* serene
As she said, "I like October,
With its splendors of decay,
But I like spring-time better,
And the warm sweet air of May."
Thus we traveled down the Switch Back,
Thus I trifled with my fate;
For she was seventeen, and I
Was turned of forty-eight.

—David L. Proudfit

A must on any Switchback rider's agenda was a visit to the burning mine at Summit Hill. The fire began deep under Slope 1 of the Lehigh Coal & Navigation Company mine on February 19, 1859. Company officials attributed the cause to "burning rubbish in an abandoned underground working," but the locals knew better. A careless workman had shoveled hot ashes from a stove into a wooden mine car, igniting the car, timber supports, and loose coal. Despite extensive efforts to control the fire, including the construction of a firewall, it worked its way westward across the valley for a mile and continued to smolder for 100 years. (Courtesy Ray Holland.)

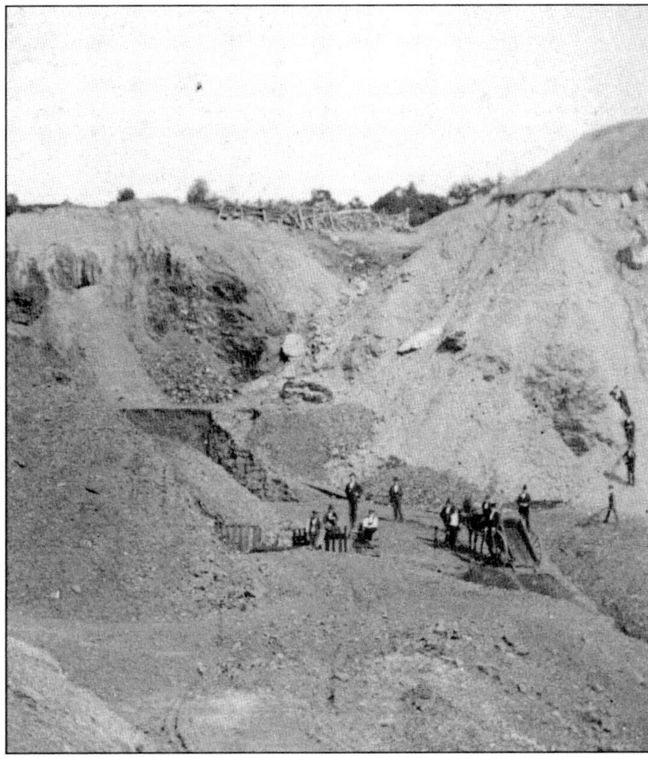

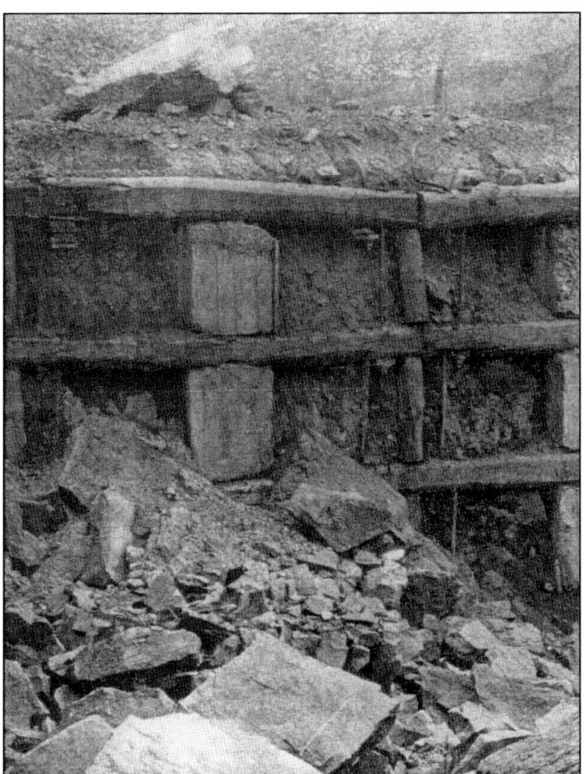

In December 1908, the Bravo Construction Company of Pittsburgh began constructing a firewall. The wall was a third of a mile long, 6 to 10 feet thick, and said to be 300 feet deep in places. At times, more than 500 men worked on the project.

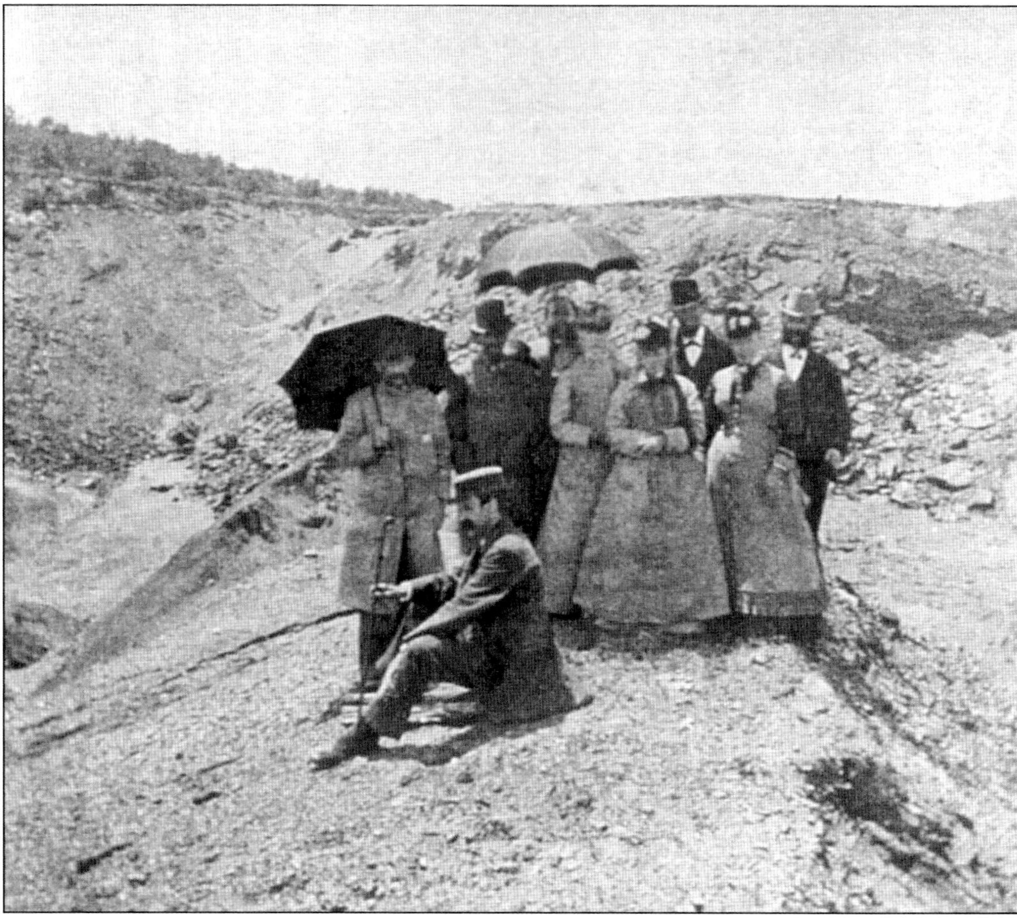

As the rattling cars of the Switchback made their way along Railroad Street in Summit Hill (today's Ludlow Street), the guide would chant, "Wanna see the burning mine?" Expecting to be confronted by Dante's inferno, ladies in long, high-necked dresses and their whiskered escorts in tailored three-piece suits were often disappointed. The only thing issuing from the ground was an odd puff of smoke. To compensate for the letdown, the guide invited visitors to put their hands over a fissure and feel the escaping heat. Convinced that a fire did indeed burn below ground, more than likely, they would reward the young man with a tip.

Two

PROSPERITY

If the first half of 19th-century Mauch Chunk belonged to Josiah White and Erskine Hazard, the second half belonged to Asa Packer. Packer was the quintessential American self-made man. Born in Mystic, Connecticut, he came to Mauch Chunk on foot in 1833, carrying carpenter's tools. Working his way up from canal laborer to boatman, boatbuilder, merchandiser, and owner of coal land and mines, he had amassed enough capital by 1855 to launch the Lehigh Valley Railroad. Mining companies no longer had to rely on slower canal transportation to carry coal to market. Packer was on his way to becoming a multimillionaire. Appointed associate judge of Carbon County in 1842, he was elected to the state legislature later that year and to the U.S. Congress in 1852. In 1866, he founded Lehigh University. An industrialist, statesman, and philanthropist, Asa Packer was America's third richest man at the time of his death in 1879. His estate was worth $54.5 million.

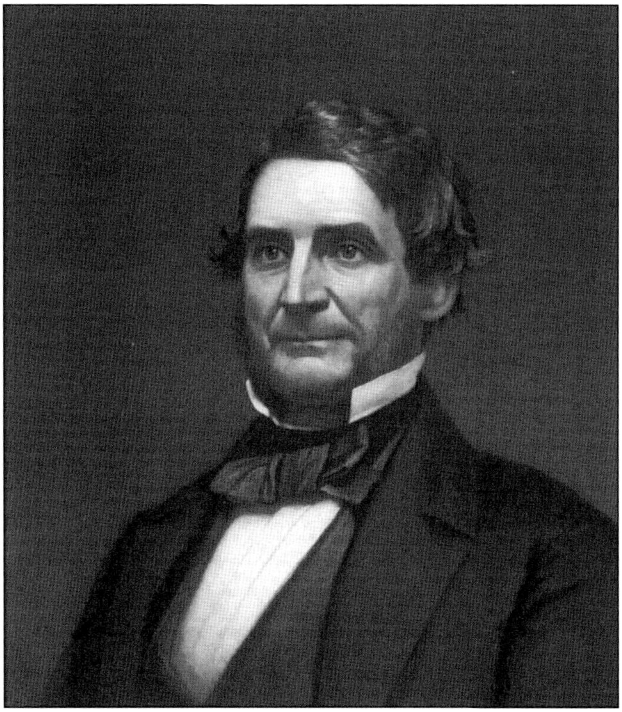

Asa Packer was a great benefactor to Mauch Chunk. In 1878, a reporter for the *Mauch Chunk Democrat* paid this tribute to the Packer family: "The Lehigh Valley was almost a wilderness when Packer and his wife arrived in Mauch Chunk but, through their efforts, is now dotted with flourishing towns, teeming with commerce and is populated with an active, industrious and generally prosperous people." When Packer died in 1879, the New York magazine *Puck* memorialized him with a cartoon. "He scattered the seeds of his benevolence with a liberal hand, and the fruits of Education and Science sprang up in his path." —From *Puck,* June 4, 1874

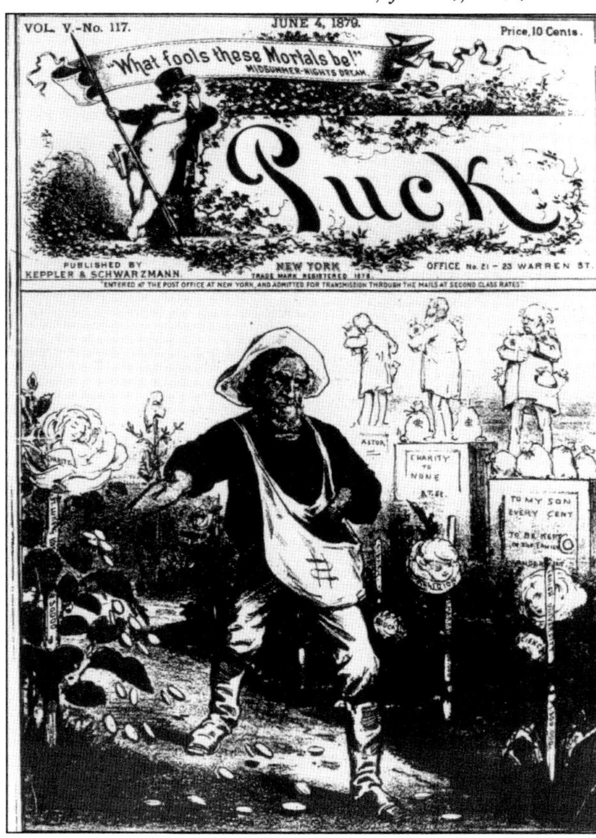

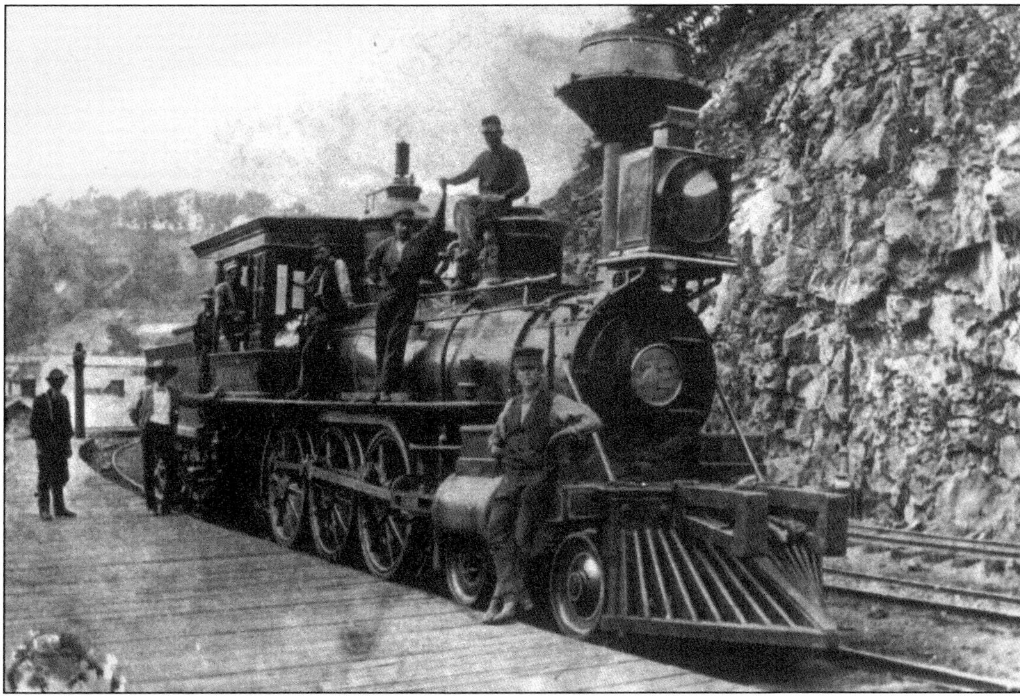

The Lehigh Valley Railroad opened on September 12, 1855, to haul anthracite from the mountains of eastern Pennsylvania to the Great Lakes and Atlantic seaboard. At its peak it had over 1,000 locomotives, a fleet of express passenger trains, a Great Lakes steamship line, and a hefty share of the Pennsylvania anthracite mining and transportation industries. Engine No. 71, the "Montana," was built in 1865 and is shown at the station platform.

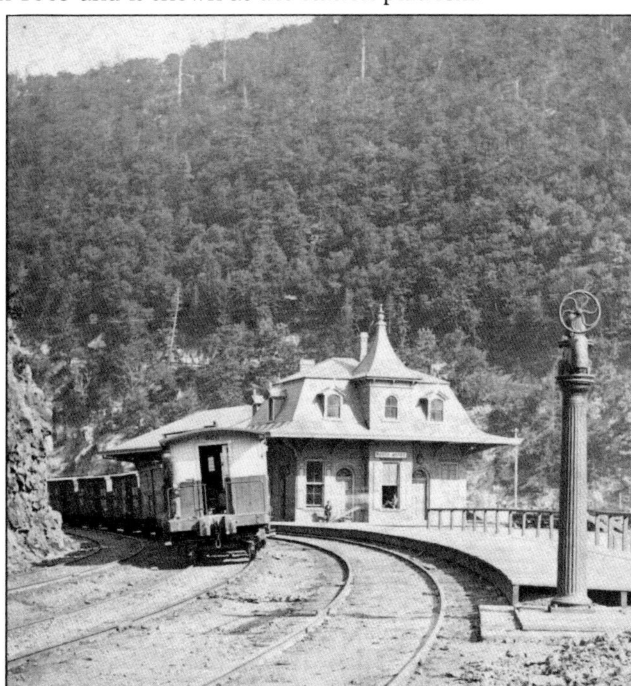

The first Lehigh Valley Railroad station was nestled at the foot of Bear Mountain. Passengers reached town by crossing a footbridge over the Lehigh River. This station was later replaced by one south of the main bridge connecting Mauch Chunk to East Mauch Chunk.

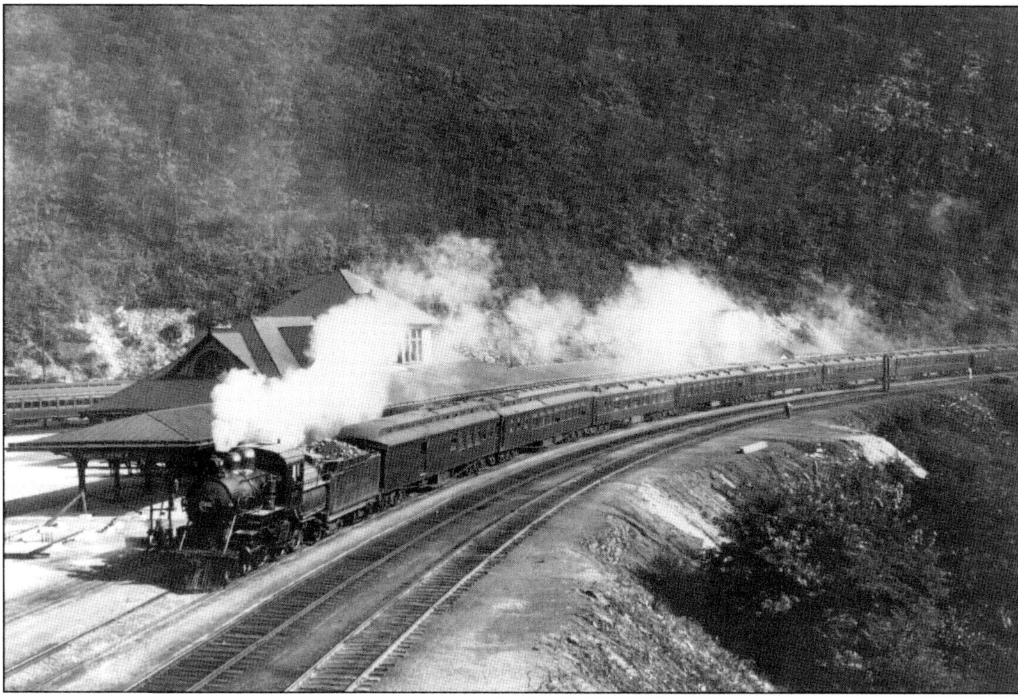

A passenger train is shown at Mauch Chunk's Second Station. (Courtesy Ray Holland/Railroad Museum of Pennsylvania.)

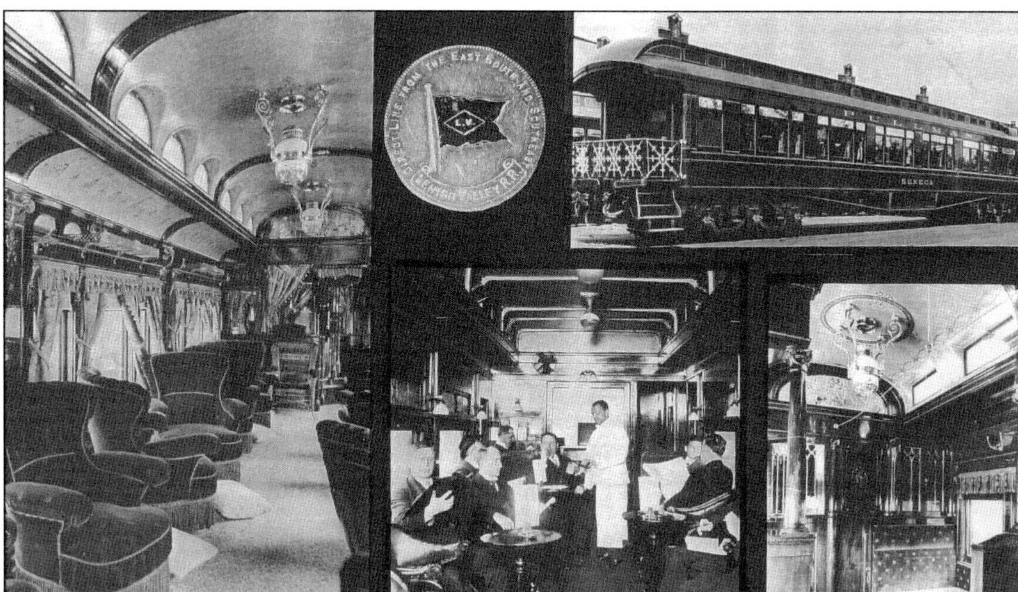

The Lehigh Valley Railroad's Black Diamond Express was known as "the most beautiful train in the world." The interior was the epitome of Victorian elegance: tasseled velvet curtains, sparkling glass-shaded ceiling lights and wall sconces, and plush swivel armchairs in the club car. In the dining car, tables were set with the finest linens, china, and silverware.

A long line of coal cars makes its way south from the coal fields. (Courtesy Ray Holland.)

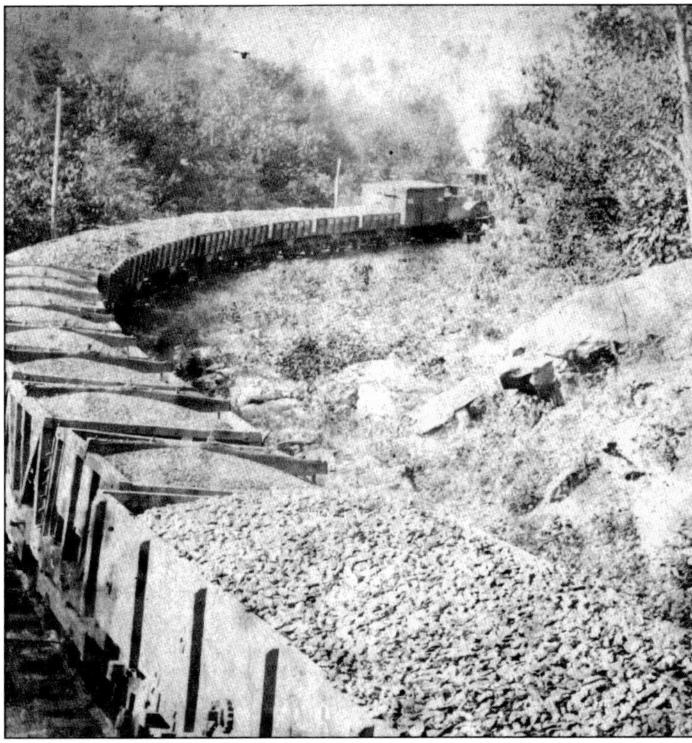

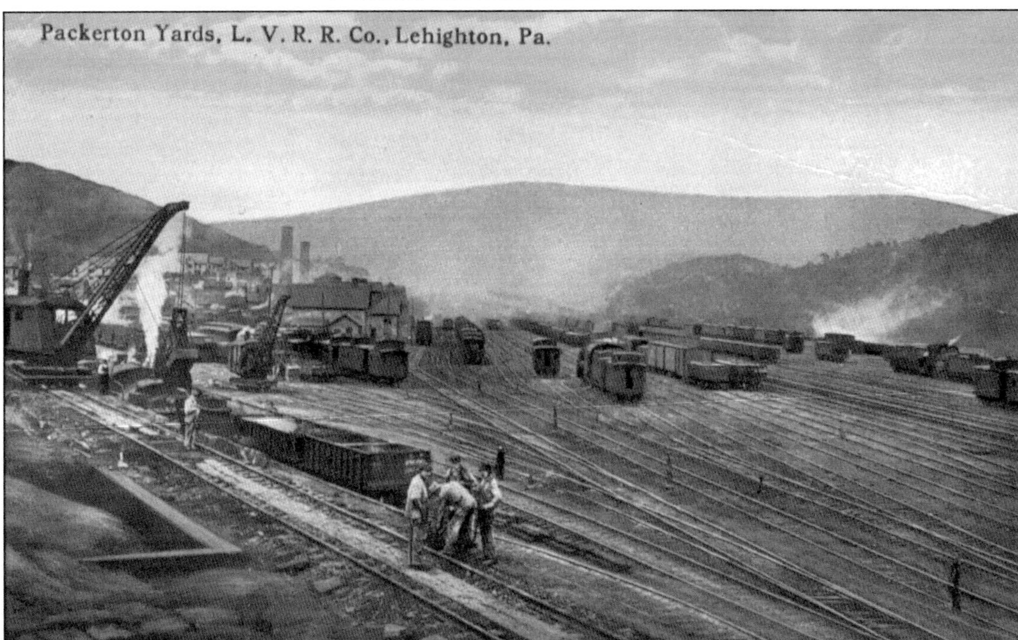

The Packerton Yards were established in 1863 on 47 acres at Burlington (renamed Packerton), midway between Mauch Chunk and Lehighton. Seven rail cars, with a capacity of more than 100 tons, could be weighed in rapid succession on the 122-foot Fairbanks scale. A train "combiner" designated coal cars for specific freight runs. Machine and car repair shops also occupied the site. By 1874, the yards employed 600 men.

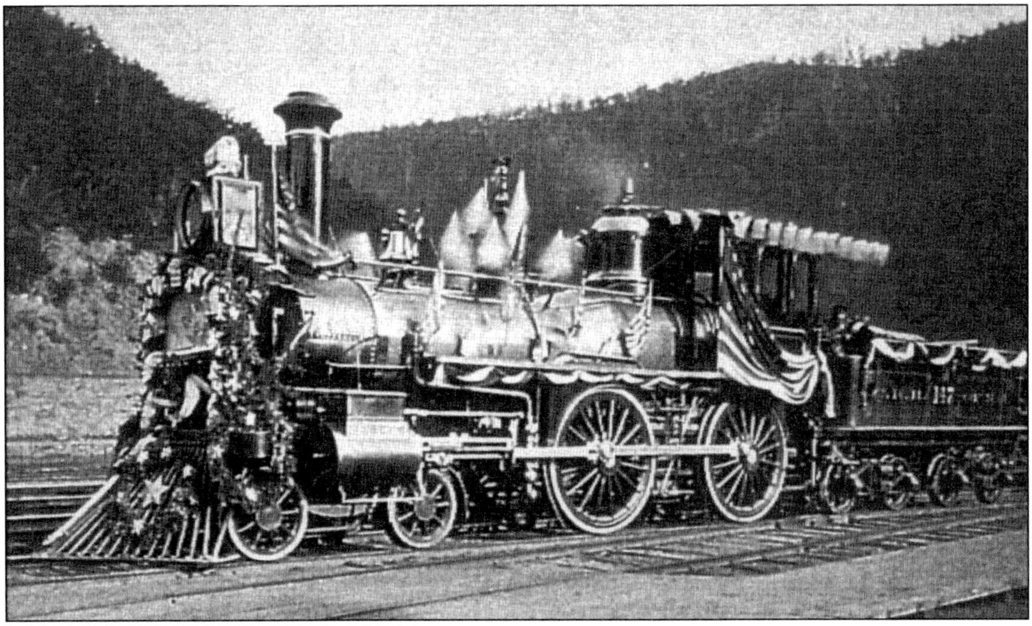

Following the flood of 1862, which destroyed the Upper Division, the Lehigh & Susquehanna Railroad, owned by the Lehigh Coal & Navigation Company, was extended from White Haven to Mauch Chunk to replace the canal. Subsequent extensions in 1867 and 1868 took the railroad to Easton and across the Delaware River to Philipsburg, New Jersey. Dead-ended in Philipsburg, the Lehigh & Susquehanna was leased by the Central Railroad of New Jersey in 1871, creating a direct route from the mines to the New York area.

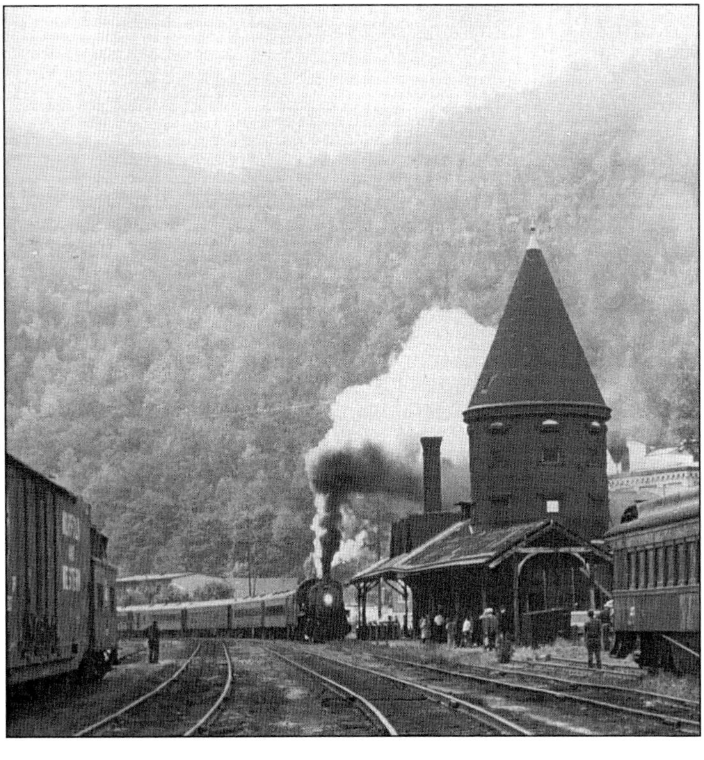

Passengers wait for the approaching train. (Courtesy Ray Holland.)

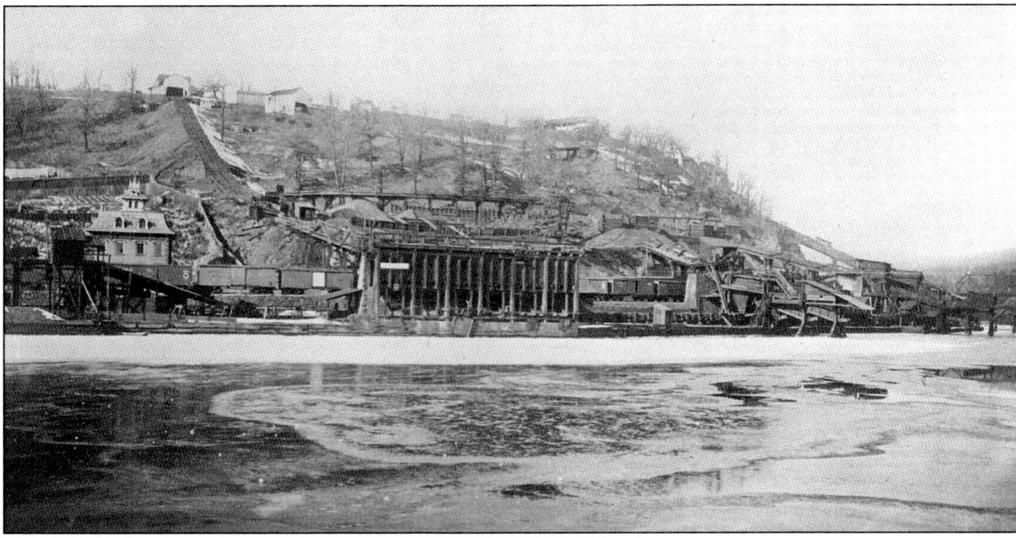

A loading facility for coal cars was on west bank of Lehigh. Switchback coal chutes are on the right of the photograph.

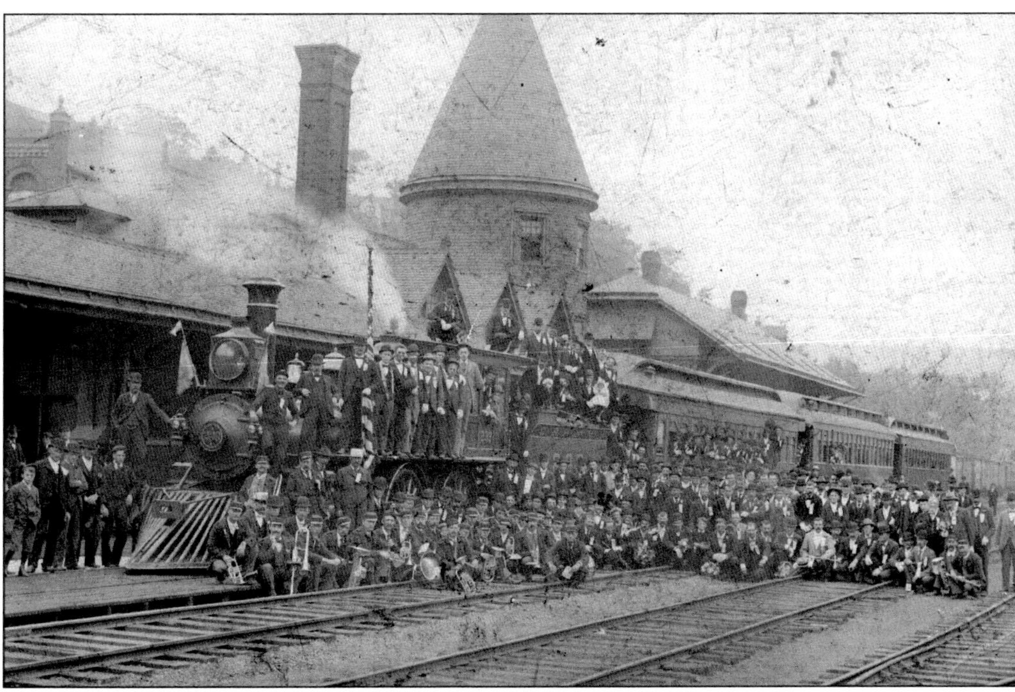

An old-time celebration at the Jersey Central Station in Mauch Chunk is shown here. (Courtesy Ray Holland.)

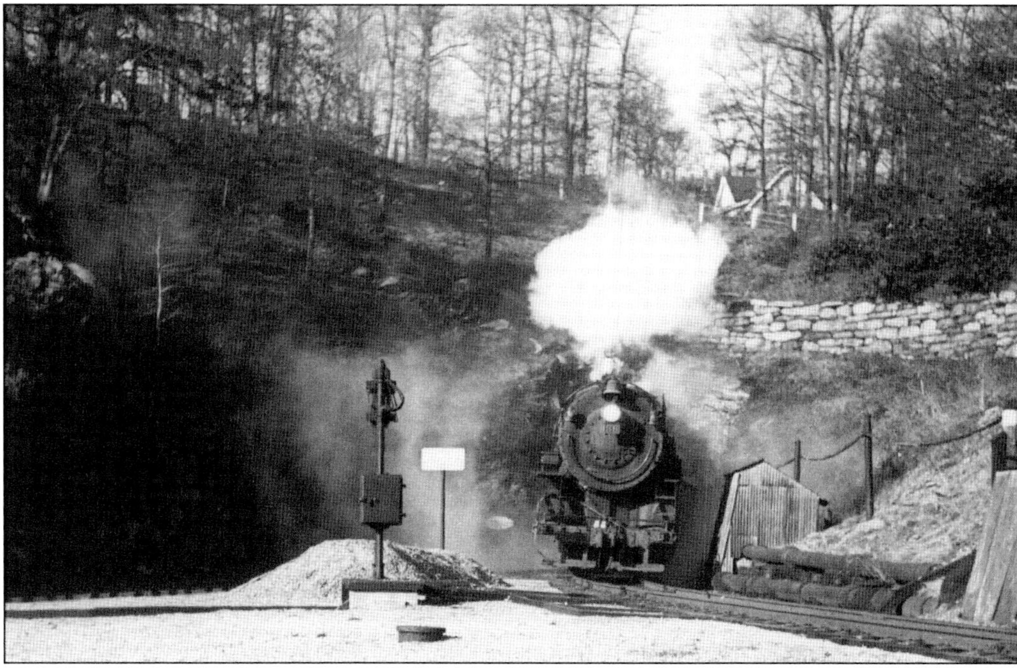

A Lehigh & New England Railroad train (owned by the Lehigh Coal & Navigation Company) exits the Hauto Tunnel. The construction of the Hauto Tunnel through Nesquehoning Mountain created a direct route from the mines and eliminated the need for the Switchback Gravity Railroad to carry coal. From 1872 until its demise in 1933, the Switchback was exclusively a passenger carrier.

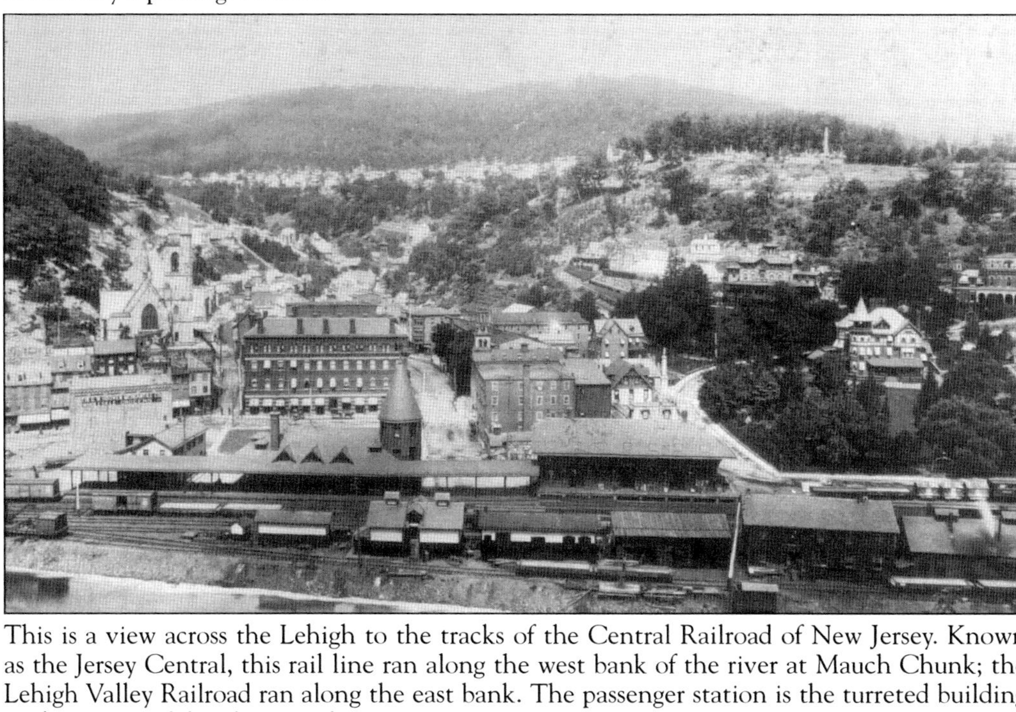

This is a view across the Lehigh to the tracks of the Central Railroad of New Jersey. Known as the Jersey Central, this rail line ran along the west bank of the river at Mauch Chunk; the Lehigh Valley Railroad ran along the east bank. The passenger station is the turreted building in the center of the photograph.

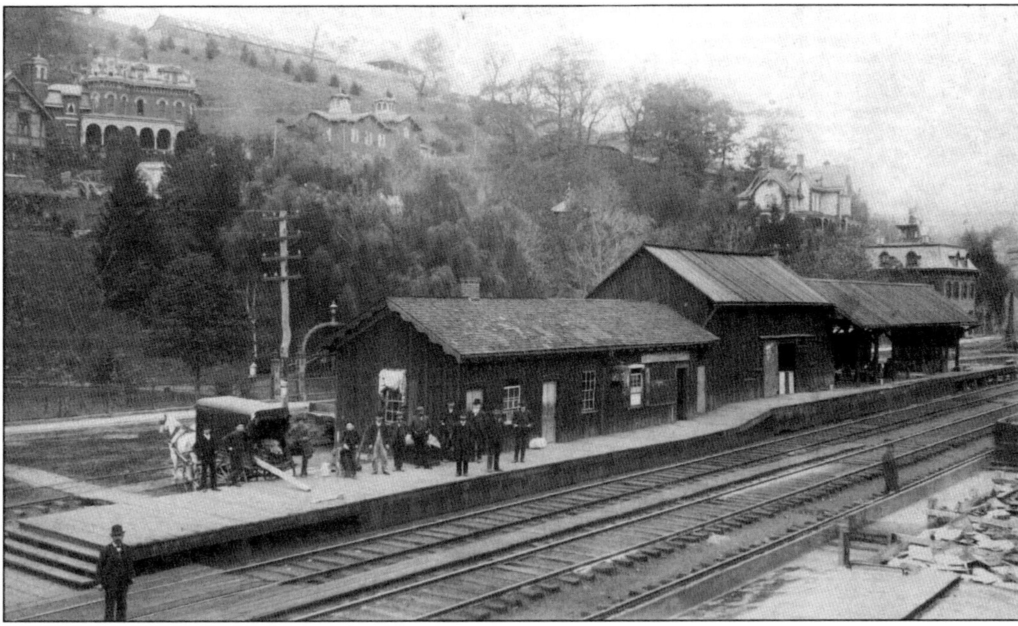

The Jersey Central freight station was located just north of the passenger station. (Courtesy Ray Holland.)

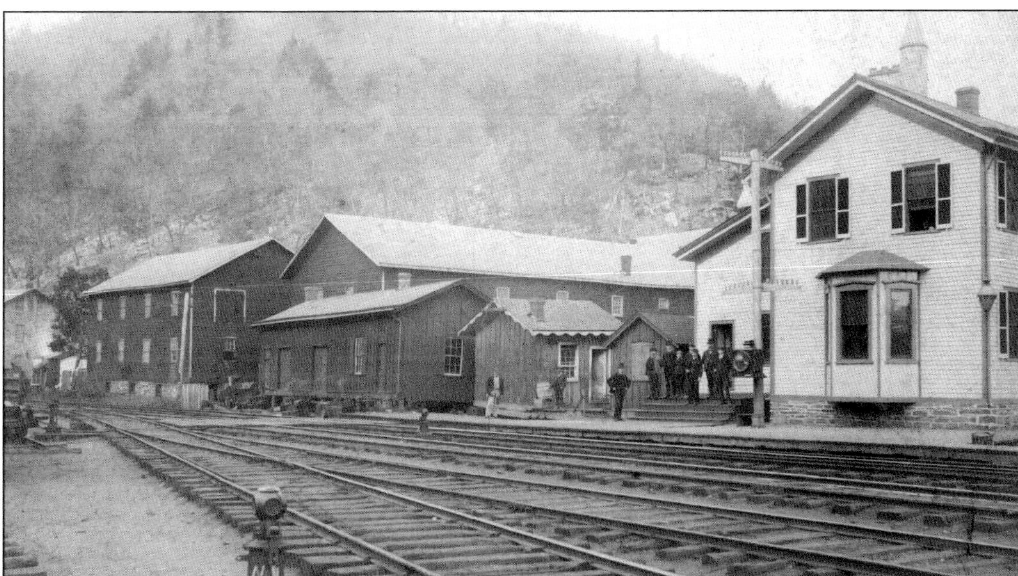

Trains were slowed down at the Jersey Central flag stop outside Mauch Chunk station. Susquehanna Street ran behind the buildings. (Courtesy Ray Holland.)

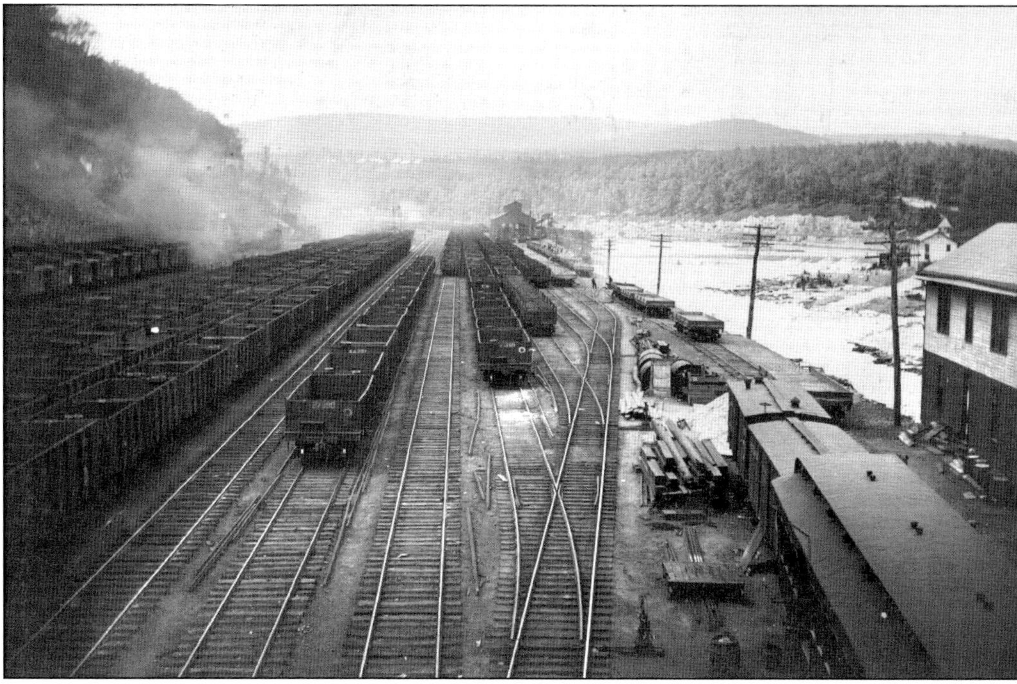

Long trains of coal cars wait to be shunted to various destinations in the Jersey Central Yards, Mauch Chunk. (Courtesy Ray Holland.)

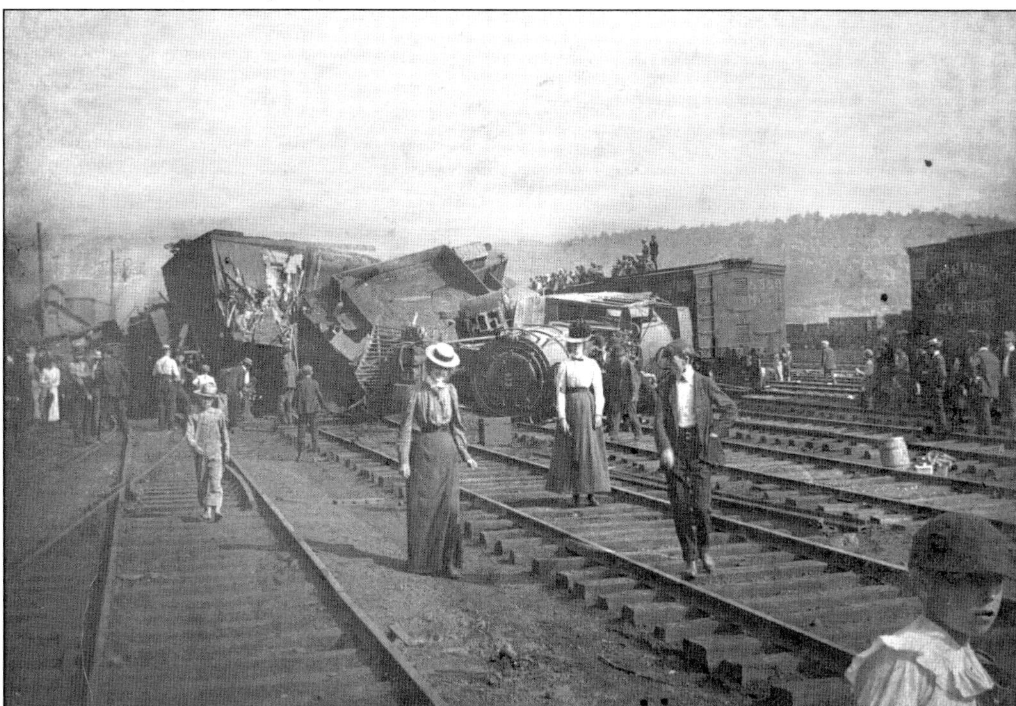

On June 26, 1891, the boiler of Jersey Central locomotive 235 exploded, killing the engineer, fireman, and two trainmen. A nearby flagman was hurt. (Courtesy Ray Holland.)

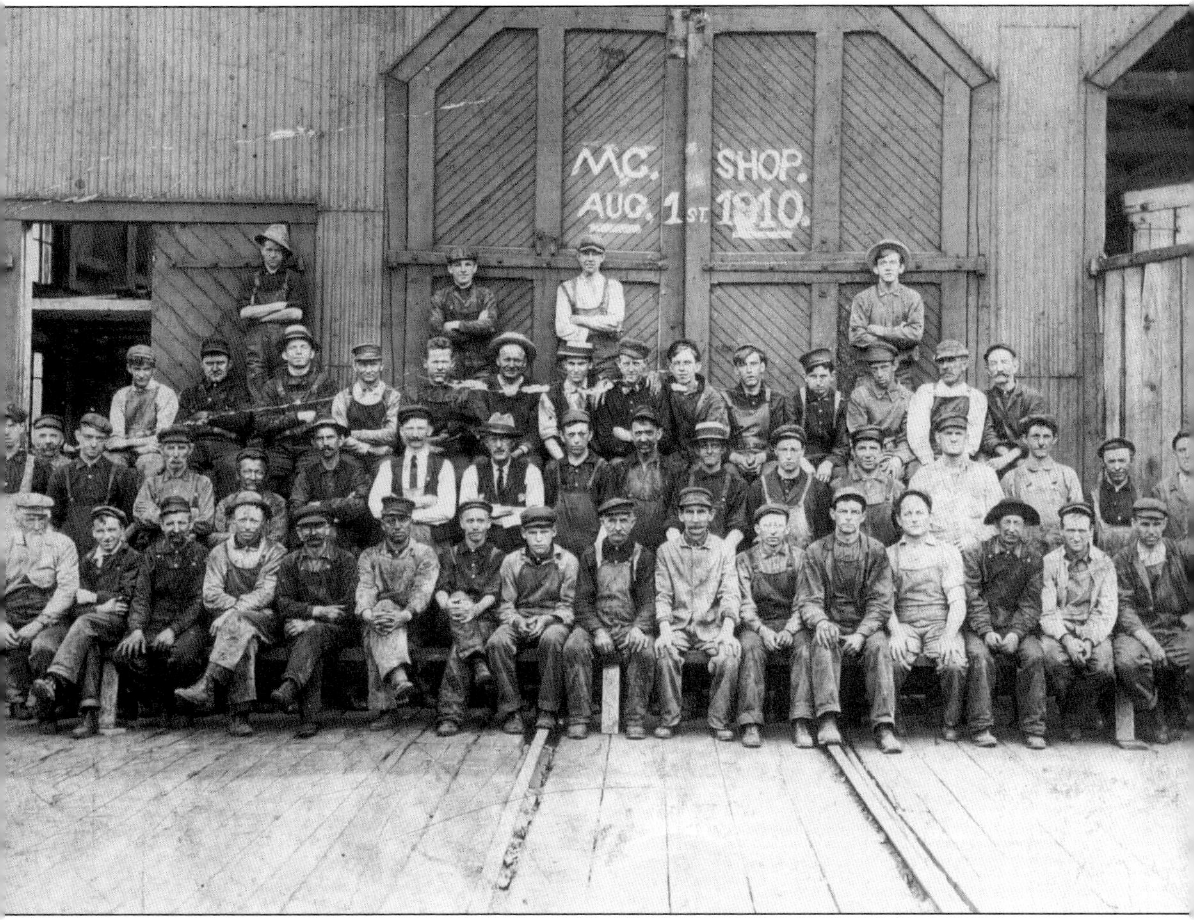

Jersey Central employees shown in this 1910 photograph are, from left to right, as follows: (first row) Mickey Lesisko, Louie Marzen, Bill Otto, John Otto, Jimmy Campbell, Morris Otto, Eddie O'Donnell, Charlie Burke, Edgar Morgan, Charles Warncke, Ed "Pops" Miller, Rube Jante, and Ben Potter; (second row) Warden Brum, Theodore Schmidt, Cy Carey, Tony Swift, Adolph Buzzard, Bill Brum, Harry Reusch, Fred Haupt, Harry Brenheiser, Wallace Eckrode, Harry Miller, Charles Herling, Bob Rimbey, Harry Haupt, Charles Reis, Patty Schlidtz McClafferty, and Clarence Fatzinger; (third row) Reinart, Butch Gruver, Charles Boetting, John Pricka, Phil Strubinger, Fred Witt, Amos Bradley, Gene Moser, Fred Eckstein, Fred Laudem, Mike "Doc" Garrah, George Long, Pete Loefflad, Ed Haupt, John Milan, and Dan McBride; (fourth row) Emory Sofranko, Vernon Batdorf, Glen Yeager, and John Brenheiser. (Courtesy Harold Queen.)

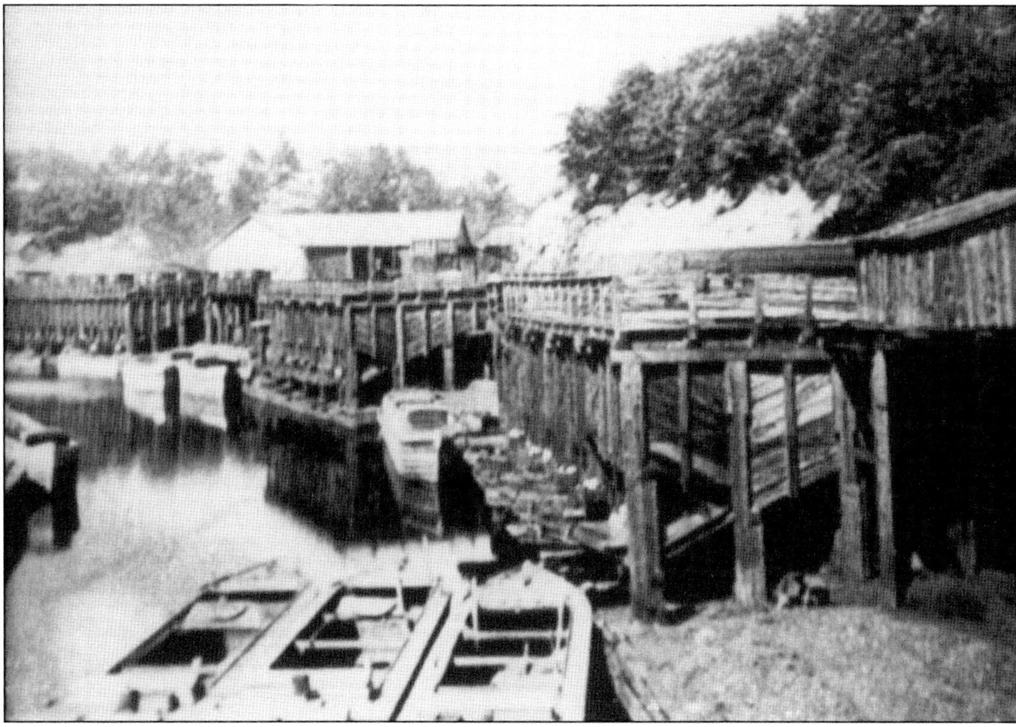

Coal boats were loaded at the Beaver Meadow Terminal in East Mauch Chunk. In 1836, long before Asa Packer connected Mauch Chunk to Easton via the Lehigh Valley Railroad, coal was hauled from the Beaver Meadow mines to Parryville, five miles south of Mauch Chunk, for shipment along the Lehigh Canal. When the disastrous flood of 1841 destroyed all the bridges between Weatherly and Parryville, East Mauch Chunk became the terminus of the railroad.

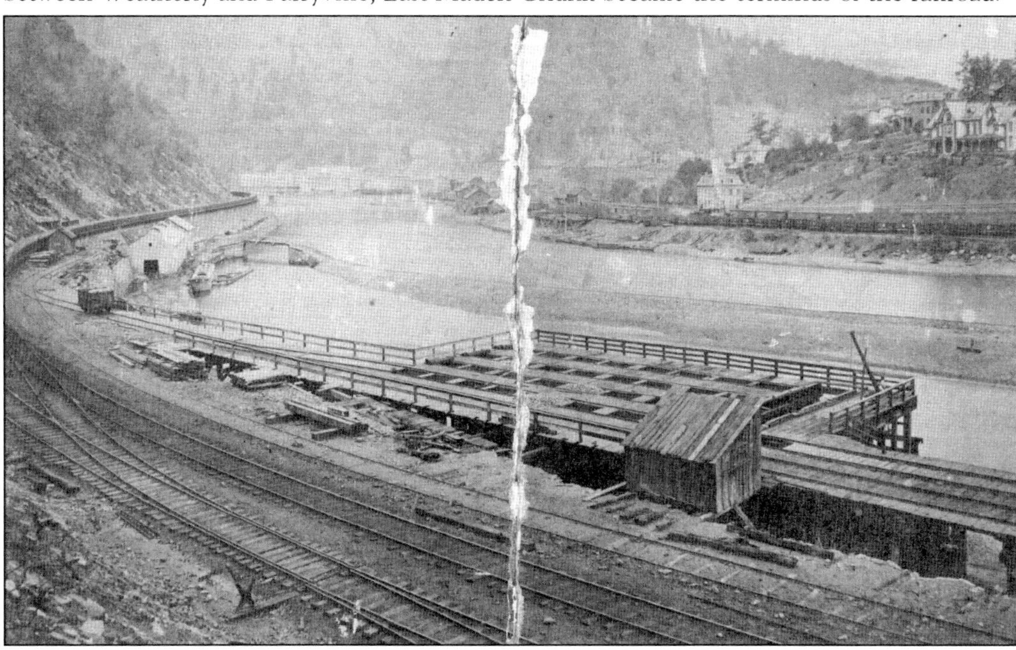

The Beaver Meadow loading docks were expanded *c.* 1900.

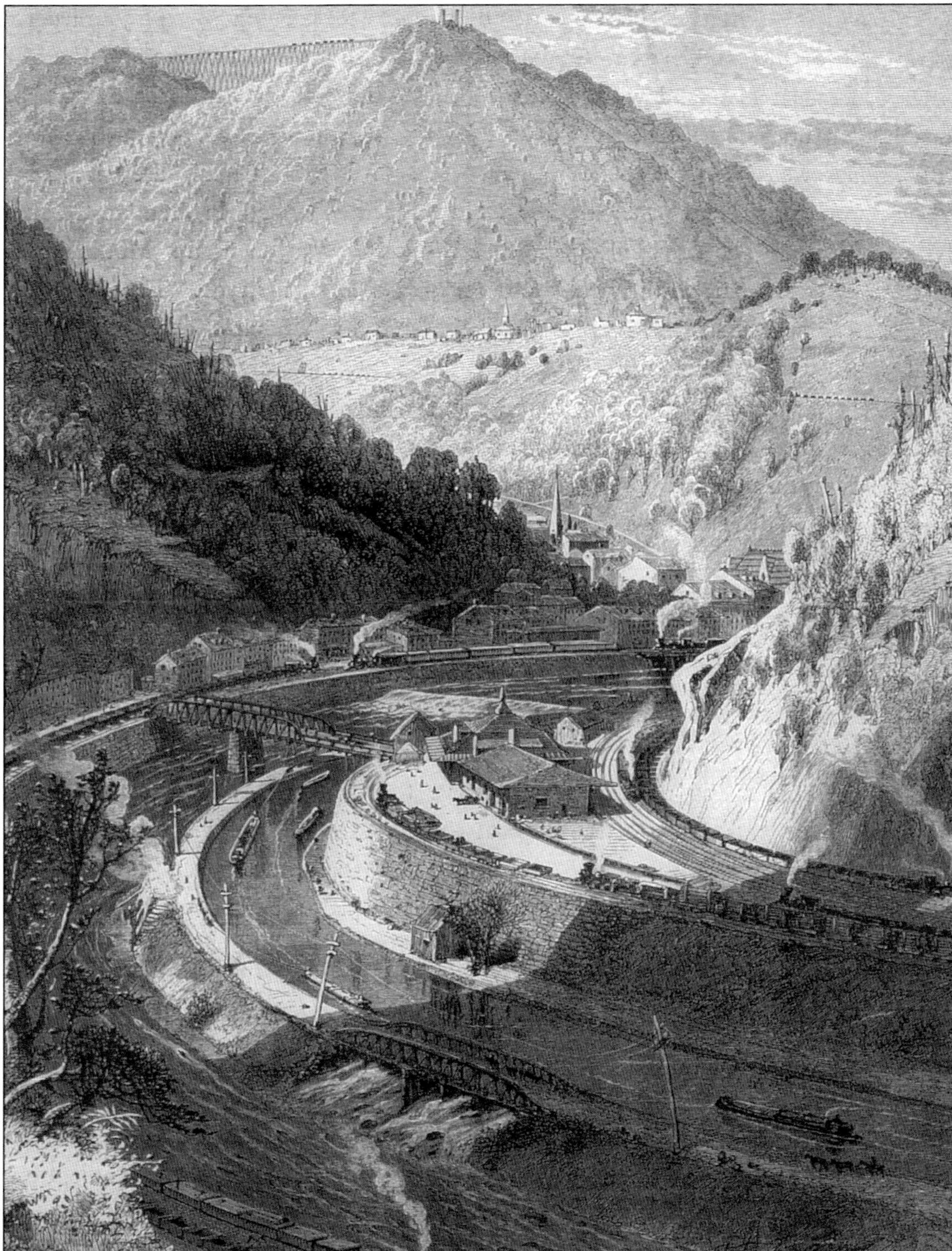

"From the town ascend the busy sound of industry, the voices of men, the whistling of steam engines, the boatman's merry horn, the rattling of coal down the chute, while in a bass accompaniment to this industrial music, there is a continuous rumbling of cars down the planes and along the level railroad." —From the diary of Walter C. Quevedo.

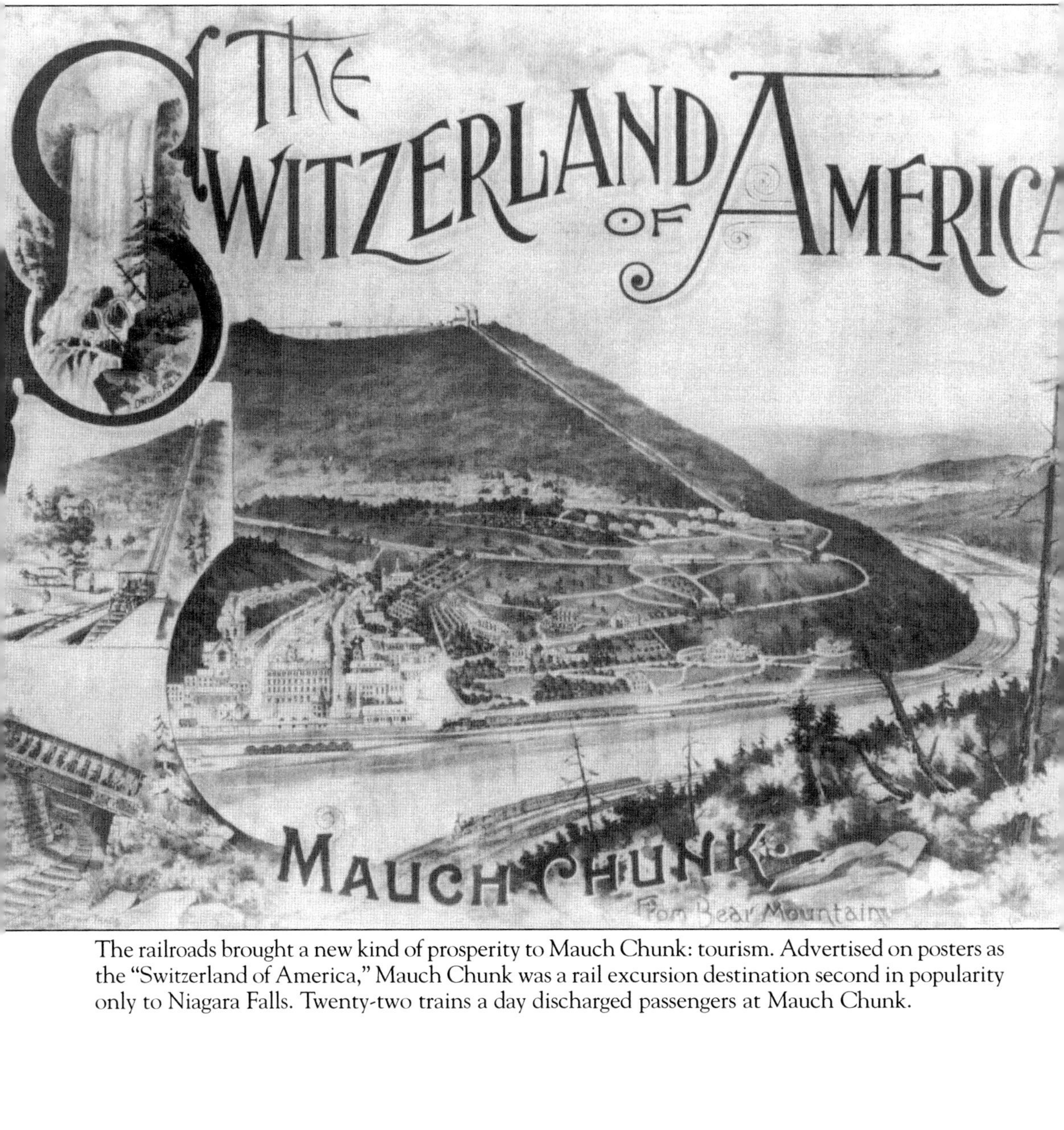

The railroads brought a new kind of prosperity to Mauch Chunk: tourism. Advertised on posters as the "Switzerland of America," Mauch Chunk was a rail excursion destination second in popularity only to Niagara Falls. Twenty-two trains a day discharged passengers at Mauch Chunk.

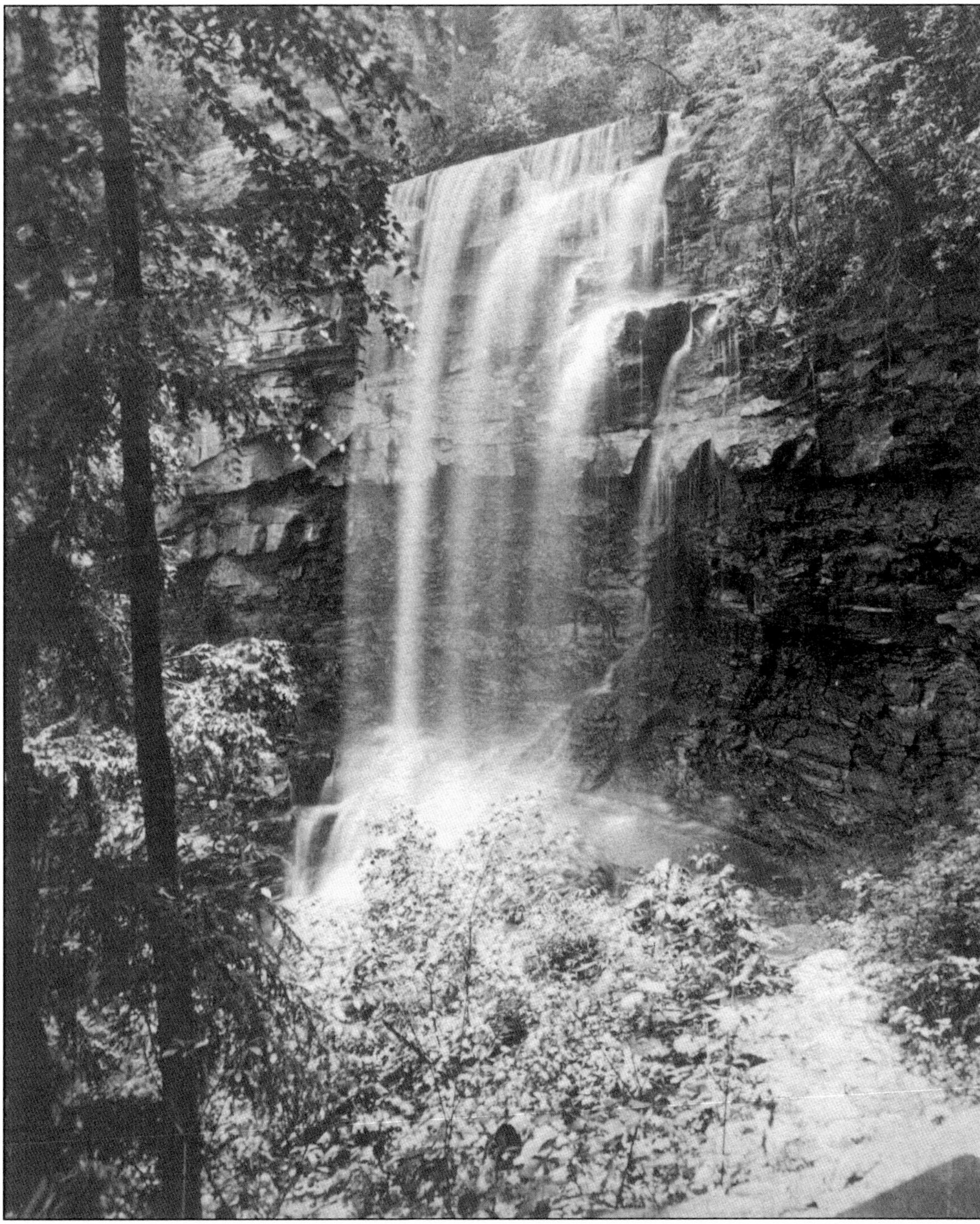

Sharing the tourist spotlight with the Switchback Gravity Railroad and serviced directly by rail in the 1870s was Glen Onoko, with its cascading stream, waterfalls, and towering cliffs. Located on the west bank of the Lehigh River (but reached through East Mauch Chunk), the Glen was a place of awe to the 19th-century tourist. The highest of the falls is the 75-foot Onoko Falls, named after the Native American princess Onoko, who leapt to her death from the summit when her father, the Lenape chief, forbade her to wed Opachee, a brave from a warring tribe.

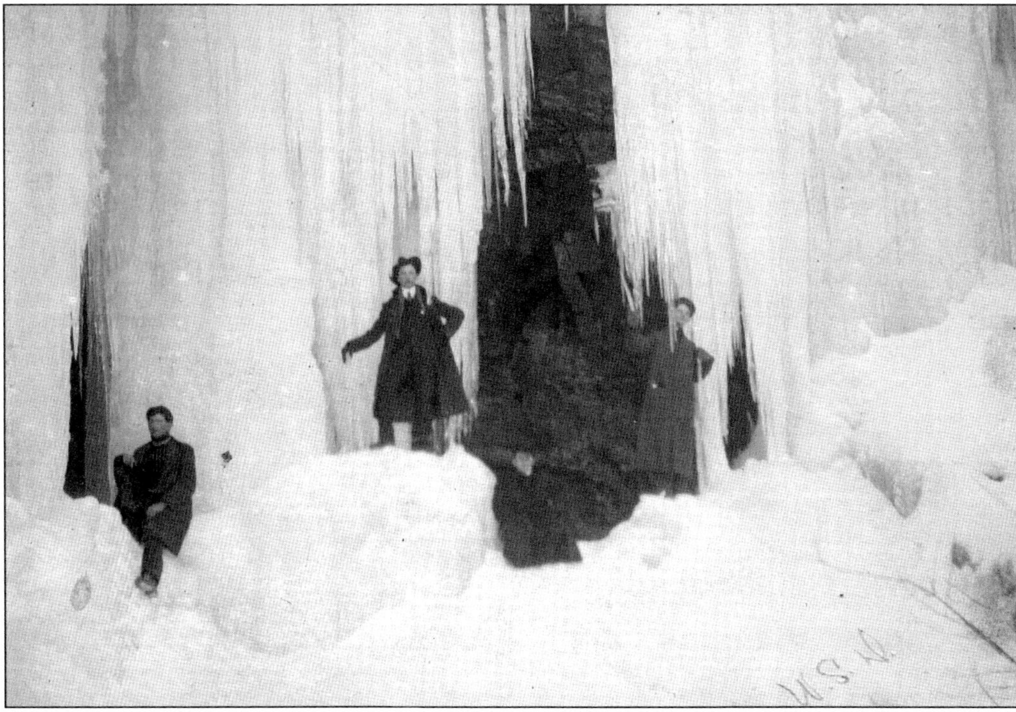

Turn-of-the-century visitors pose in front of the ice-bearded Falls of Onoko.

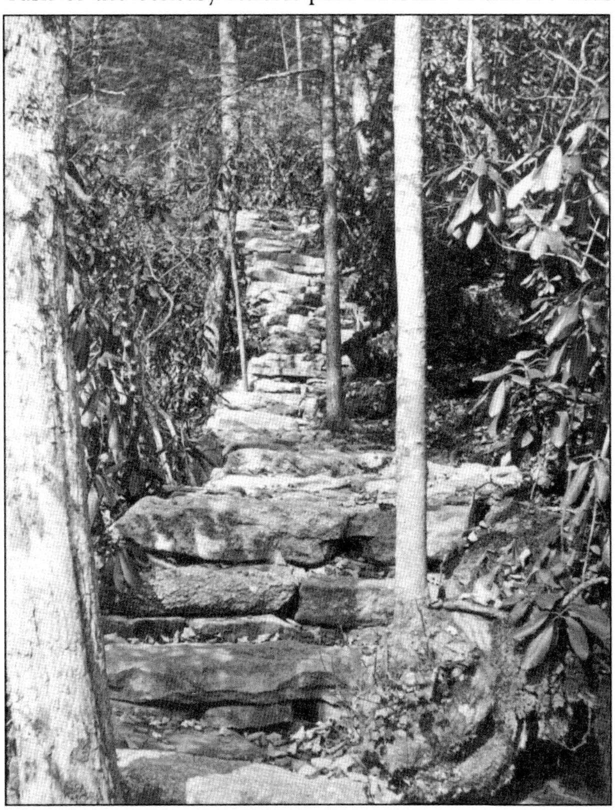

Names like Hidden Sweet, Crystal Terrace Cascades, Moss Cascade, Lover's Bath, Pulpit Rocks, Spectre Cascade, Dual Vista, Heart of the Glen, Chameleon Falls, Falls of Onoko, Terrace Cascade, Cave Falls, Home of the Mist, and Sunrise Point characterize the romanticism and charm of cascade, waterfall, and rock formation. Ascent to the falls was made easy in the late 19th century. The Wahnetah Land & Improvement Company built stone steps up the hillside and bridges across the cascades. Today, the climber is on her or his own.

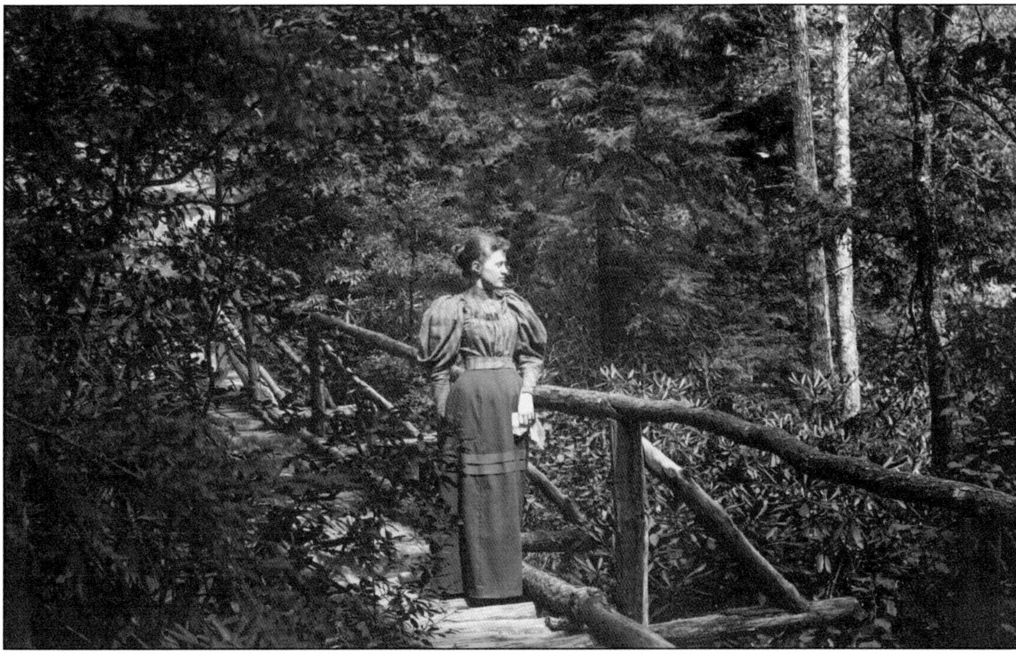

The trail leads upward through a chasm canopied by evergreen and deciduous trees and made dark and lush by wild rhododendron and mountain laurel. The roar of the falls grows louder and, suddenly, the magical cataracts are there, churning up cool dampness as white water cascades, foams, splashes, and sparkles over ledge and rock.

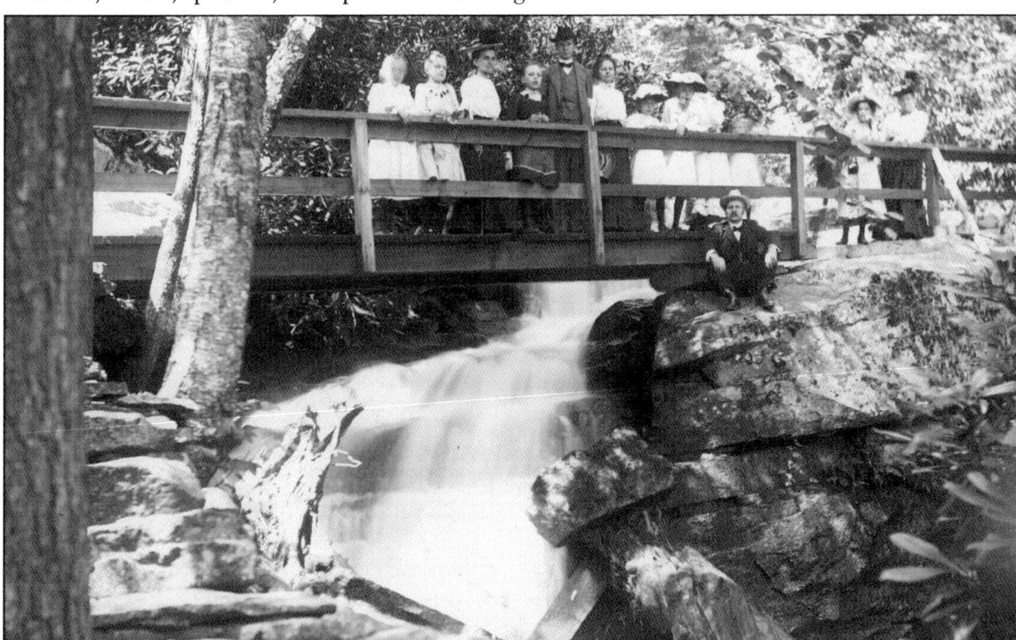

"Here and there, from between clefts in the mountainside, the pure, sparkling, God-given cold water springs forth as if seeking the sunlight in which it shimmers and glistens, throwing its golden spray in all directions, and beautifying the scene with a myriad miniature rainbows." —From the diary of Walter C. Quevedo. (Courtesy Ray Holland.)

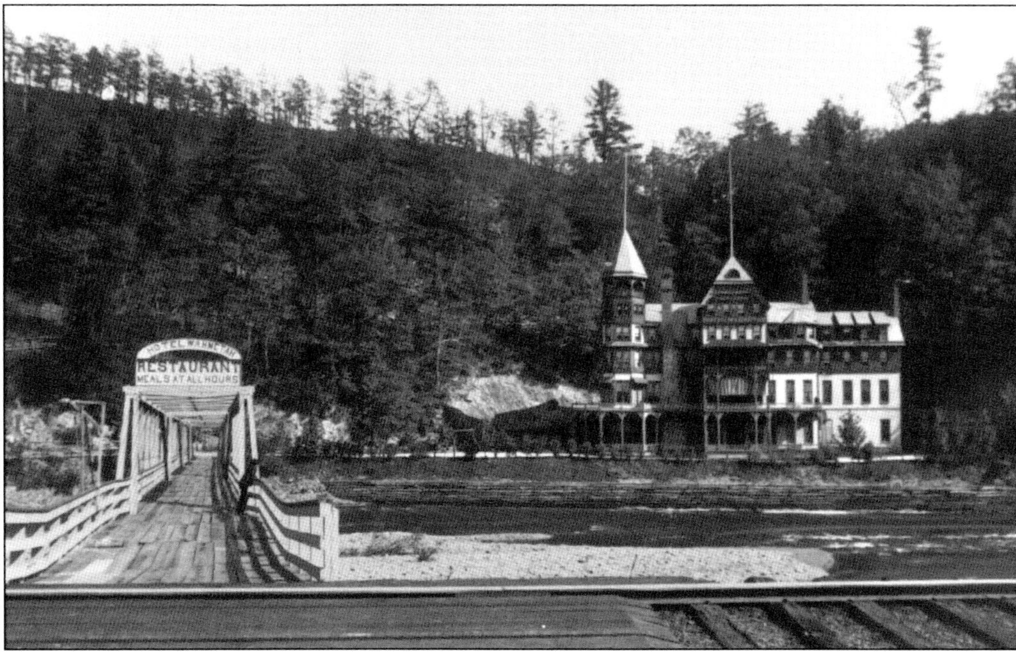

Built in 1885, the Onoko Tavern was such a popular watering hole for visitors to the falls that it was enlarged to 47 rooms and renamed the Wahnetah, after a local Native American princess. With a large dance pavilion, 84-foot bar, horse-and-carriage rides, tennis courts, and scenic trails up the falls, it was a veritable 19th-century Club Med. On April 27, 1911, a raging forest fire sparked the famous landmark. Two hours later, it was a smoldering ruin.

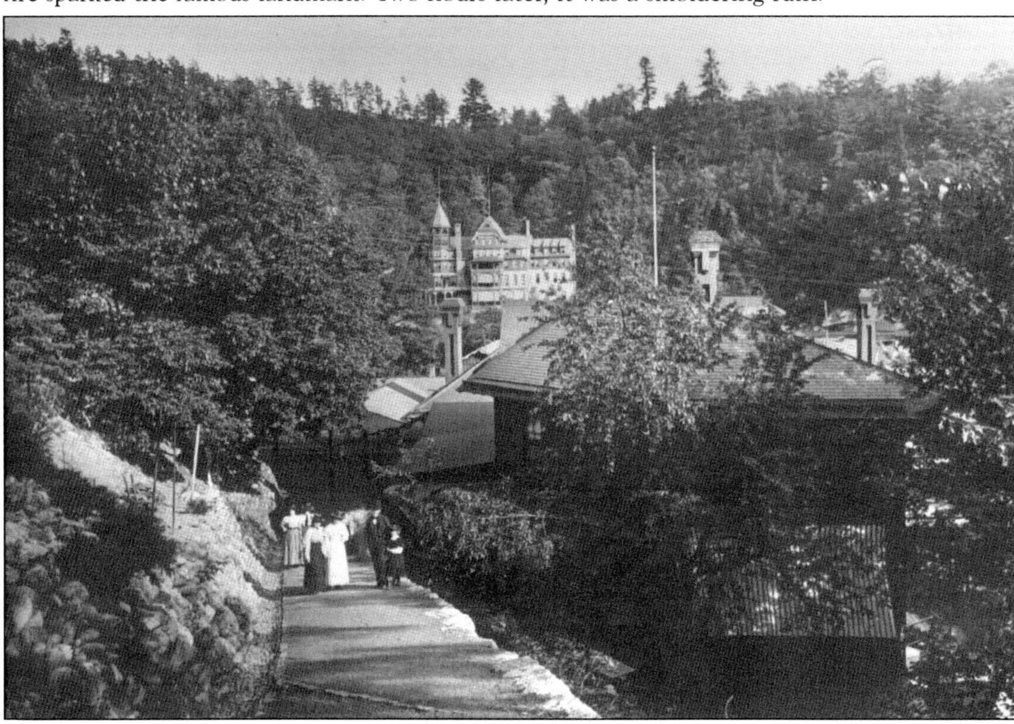

Tourists, just off the train, begin the climb to the falls.

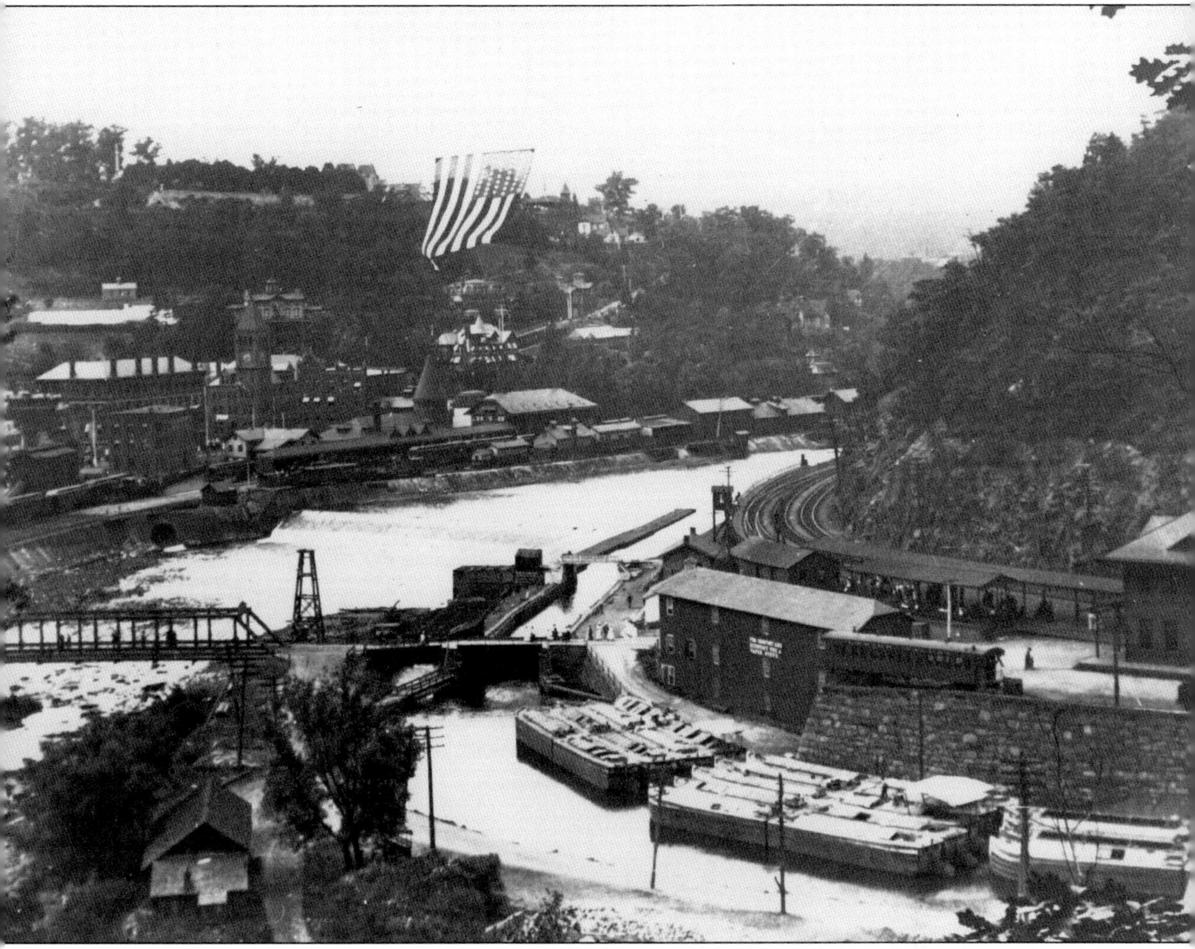

Once a lookout for Lenape Indians, Flagstaff Mountain is 1,600 feet above sea level and almost 1,000 feet above town. The mountain got its name at the start of the Civil War when a group of patriotic Unionists flew the American flag from the top of a dead hemlock tree on its summit. Mauch Chunk's most famous Fourth of July celebration took place in 1898, when rail excursion tourism was at its peak. Following the sinking of the battleship *Maine* in Havana Harbor, precipitating the Spanish-American War, the largest American flag ever flown to that date was unfurled across the Lehigh River between Flagstaff and Bear Mountains. The flag measured 75 feet, 8 inches long by 57 feet wide and weighed 265 pounds. Ten thousand spectators cheered, cannons boomed, whistles blew, and bands played as the enormous flag moved from Bear Mountain across the Lehigh on its 3,000-foot cable. (Courtesy Ray Holland.)

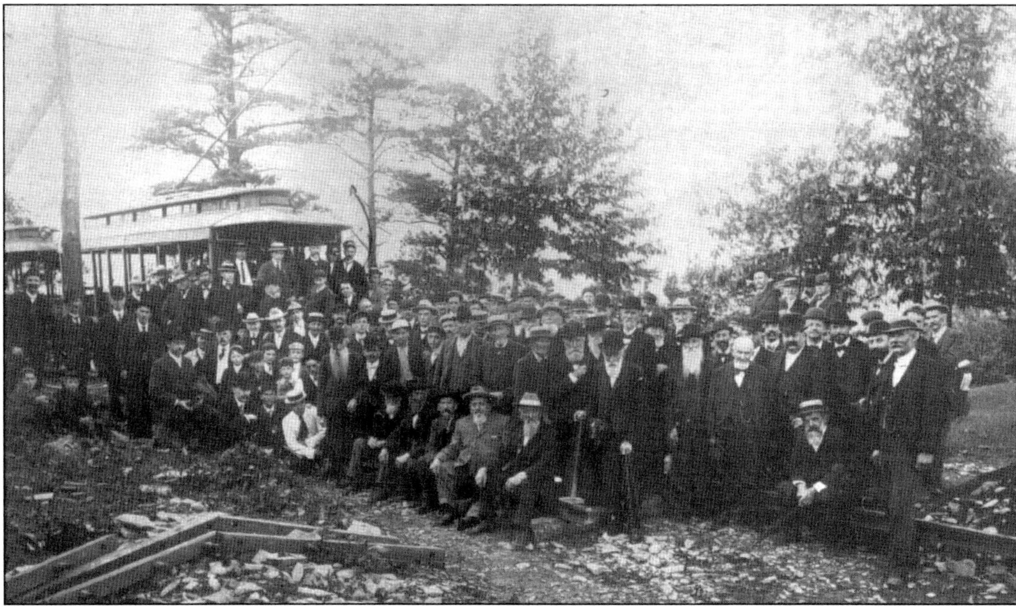

Flagstaff Park was established in 1901, but it was not a popular tourist destination until 1903, when the Mauch Chunk, Lehighton, & Slatington Railway Company (subsequently the Carbon Transit Company) inaugurated trolley service from Mauch Chunk to Lehighton via Flagstaff Mountain. Here, townspeople and dignitaries attend the groundbreaking celebration.

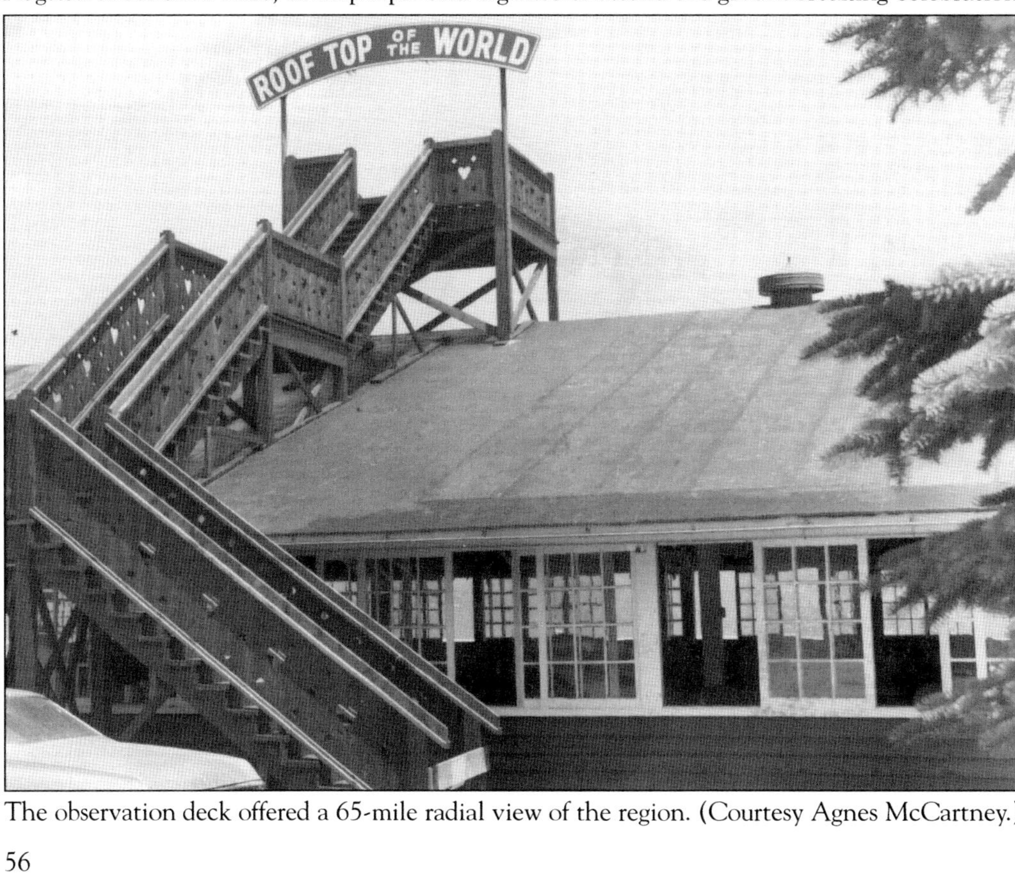

The observation deck offered a 65-mile radial view of the region. (Courtesy Agnes McCartney.)

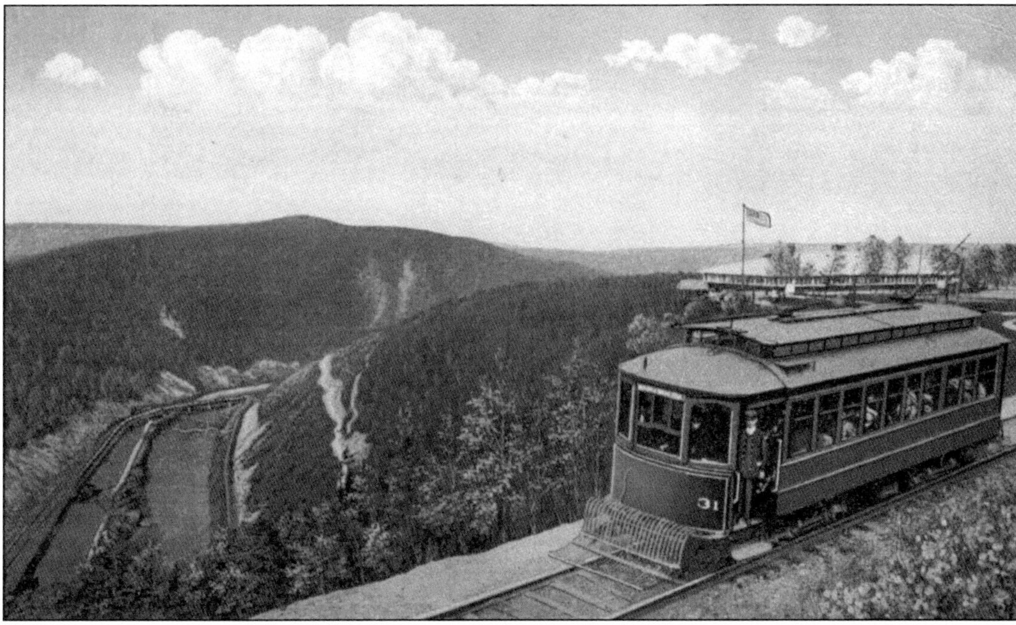

Trolley riders were afforded a panoramic view of Mauch Chunk and the Lehigh River Valley as they rode to the top of Flagstaff Mountain.

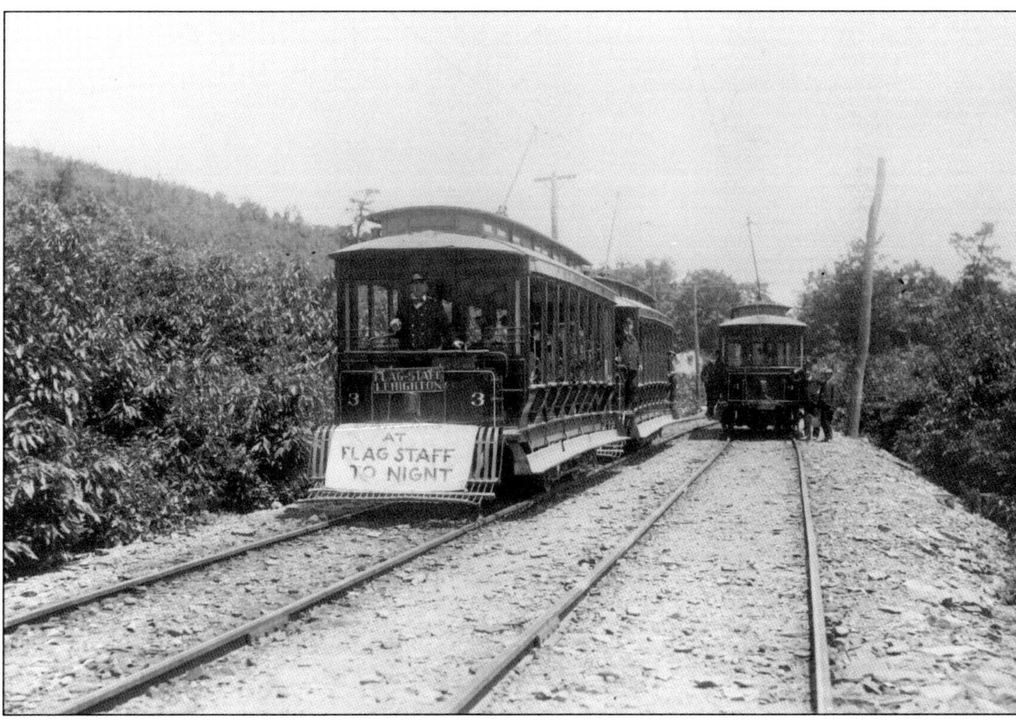

Three trolleys stop on the grade. As the sign on the front trolley indicates, entertainment at Flagstaff did not stop at dark. Amusement park rides and the famous dance pavilion were open well into the night.

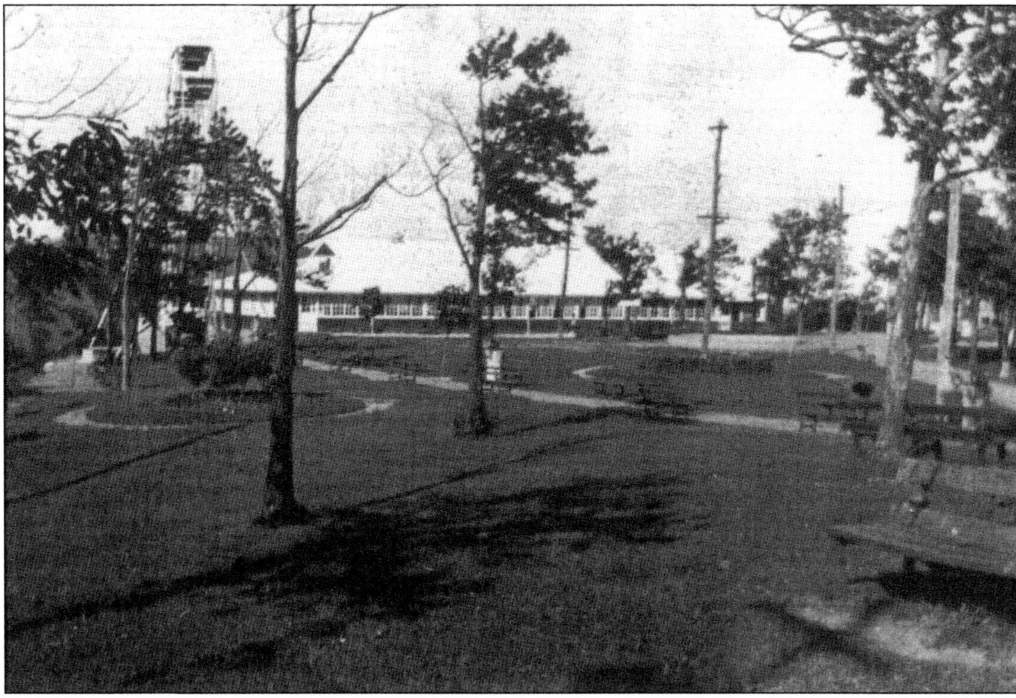

The new dance pavilion, known as the Ballroom in the Clouds, opened in 1908 on Flagstaff. The famous Ferris wheel can be seen to the left.

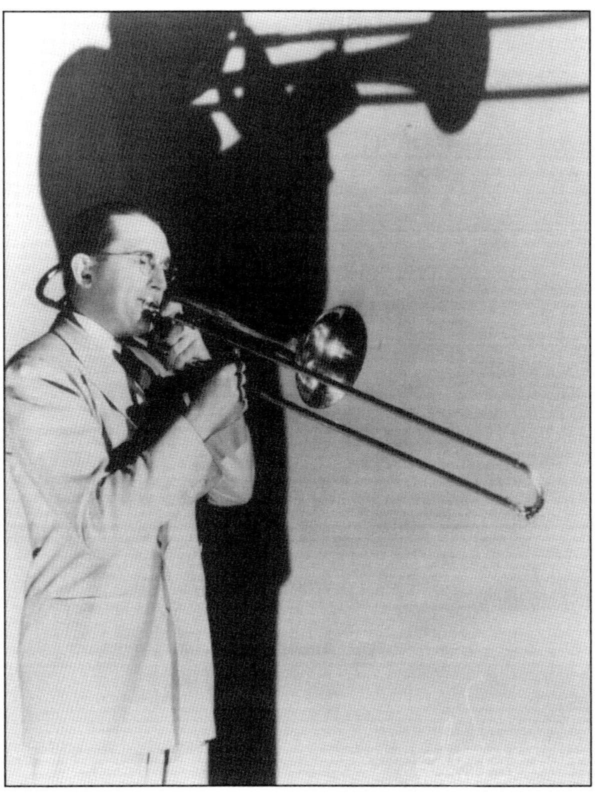

Tommy Dorsey plays his trombone in a publicity photograph. The Dorsey Brothers Band, from the neighboring town of Lansford, often played at the Ballroom in the Clouds. (Courtesy Richie Lisella.)

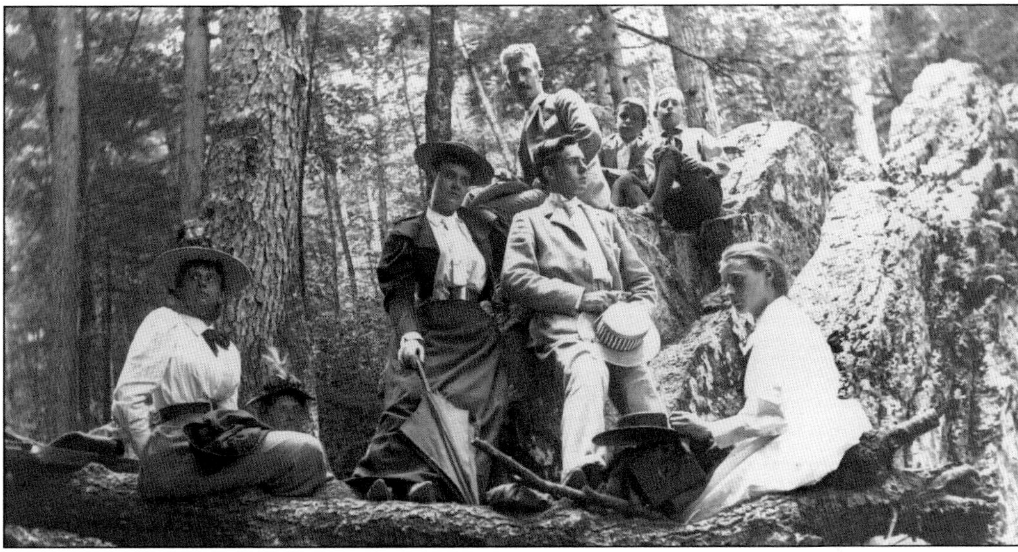

Picnickers hold still for a long exposure. In the golden age of tourism, Mauch Chunk had a number of notable visitors, including Henry Wadsworth Longfellow, John James Audubon, Charles Dickens, Ulysses S. Grant, Louis Comfort Tiffany, and Thomas Edison. Alice Austen, documentary photographer, came to Mauch Chunk in 1893 with a group of friends. She recorded her stay on glass slides. Her photographs, covering more than 50 years of social history, may be viewed at the Alice Austen House Museum, 2 Hylan Boulevard, Staten Island, New York. (Courtesy Staten Island Historical Society, Alice Austen House Collections.)

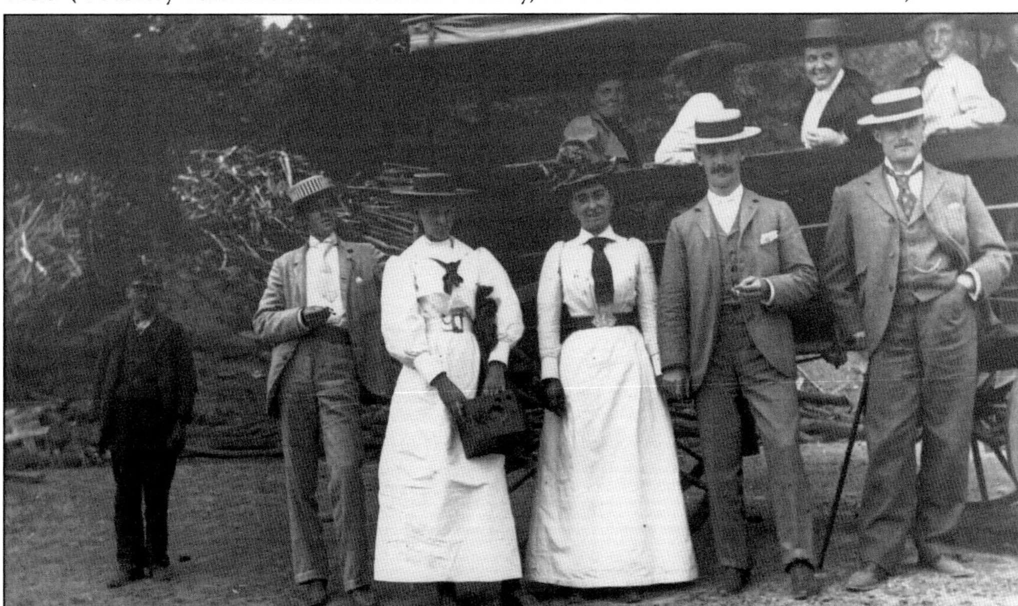

This photograph is titled "Group at Tunnel Mine Entrance," by Alice Austen. The mine is believed to be the Hacklebernie Mine on Mount Pisgah. By 1825, a 300-foot shaft had been bored into the mountain two and a half miles from the Lehigh River. No less than 20 different contractors took coal from this Lehigh Coal & Navigation Company mine over a period of 60 years. (Courtesy Staten Island Historical Society, Alice Austen House Collections.)

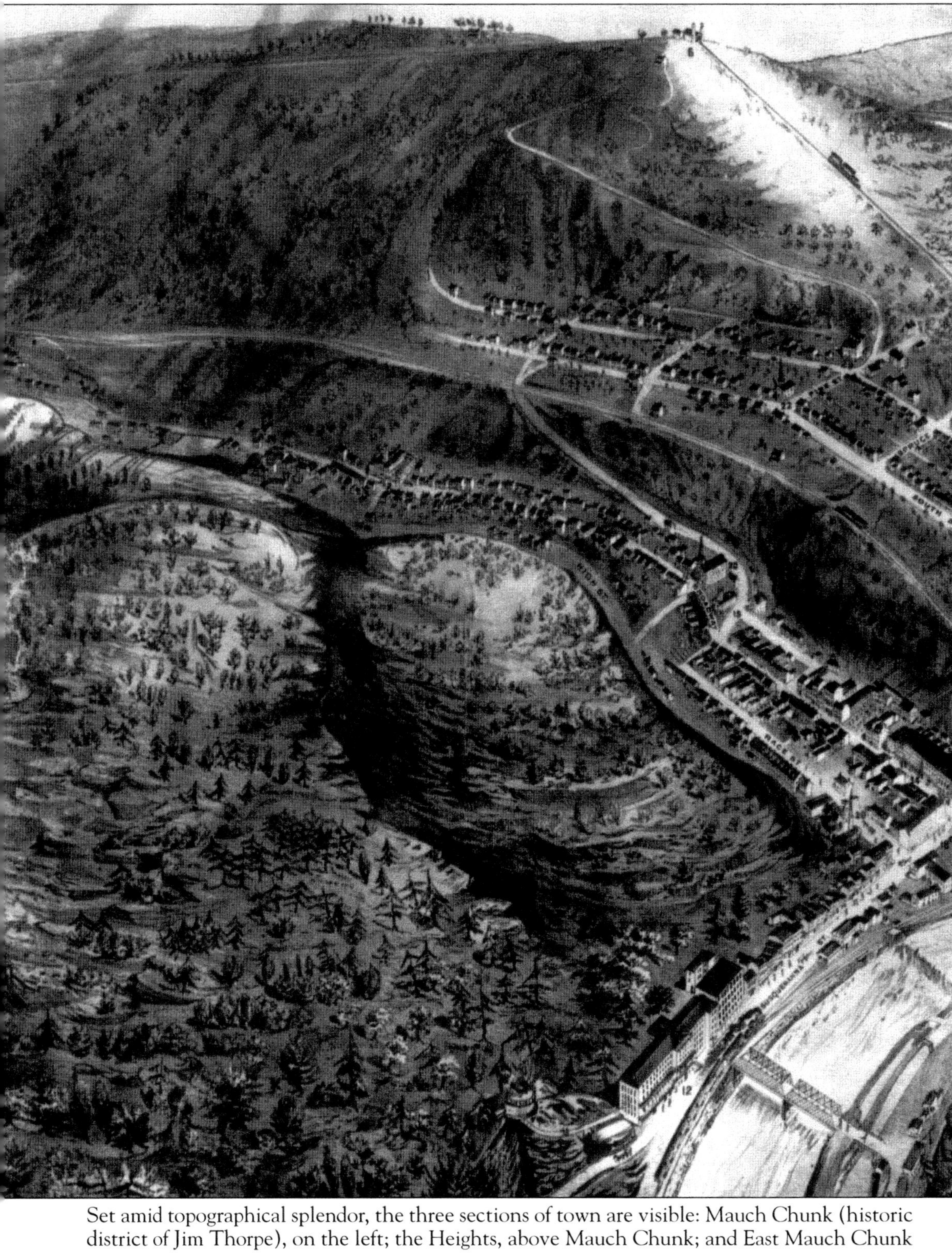

Set amid topographical splendor, the three sections of town are visible: Mauch Chunk (historic district of Jim Thorpe), on the left; the Heights, above Mauch Chunk; and East Mauch Chunk

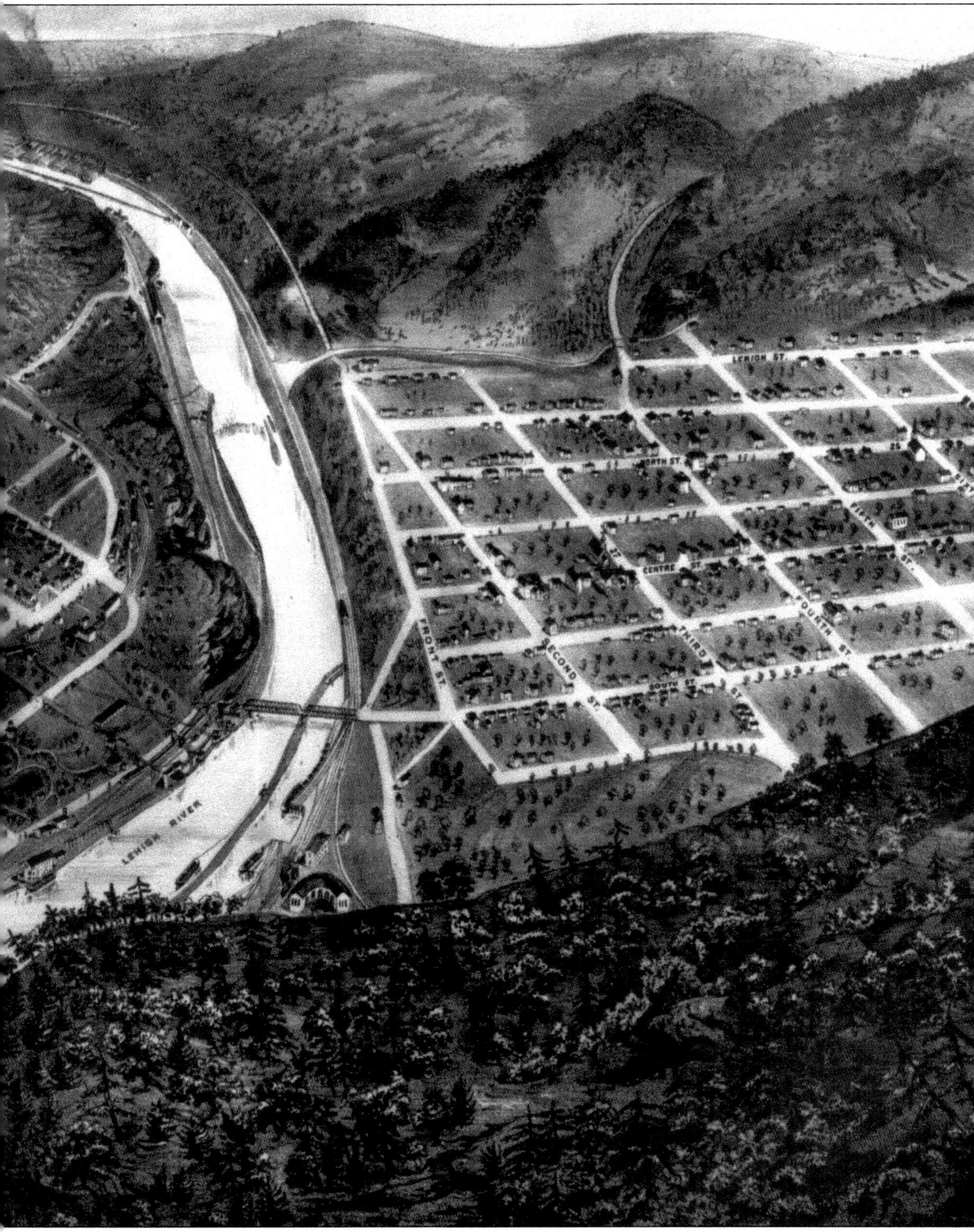

(East Jim Thorpe), across the Lehigh River.

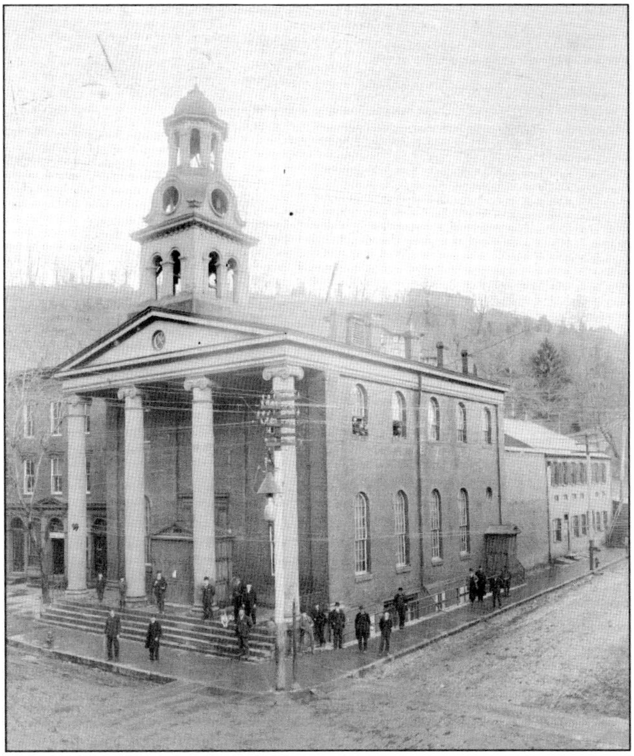

Mauch Chunk's second courthouse was built after the great fire of 1849. By the second half of the 19th century, anthracite was the principal fuel of the new industrial age. As the demand for coal grew, so did Mauch Chunk's prosperity. Imposing stone-and-brick buildings befitting the seat of the newly formed Carbon County transformed the coal-shipping outpost into an elegant Victorian town.

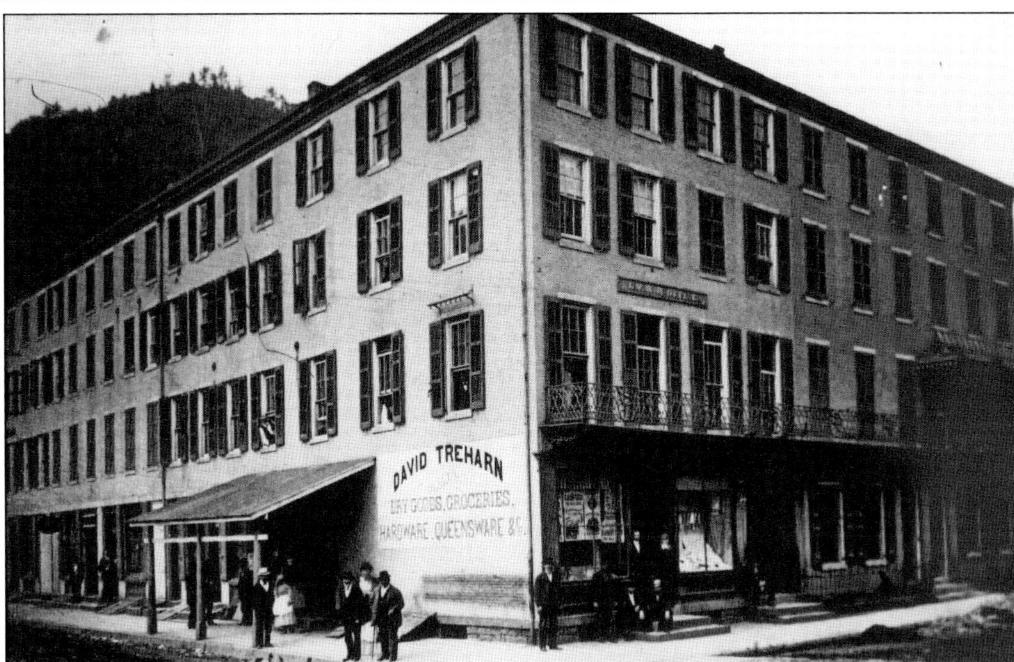

This commercial building, opposite the courthouse, housed the offices of the Lehigh Valley Railroad. It was torn down and replaced by the present Lehigh Coal & Navigation Building in 1882.

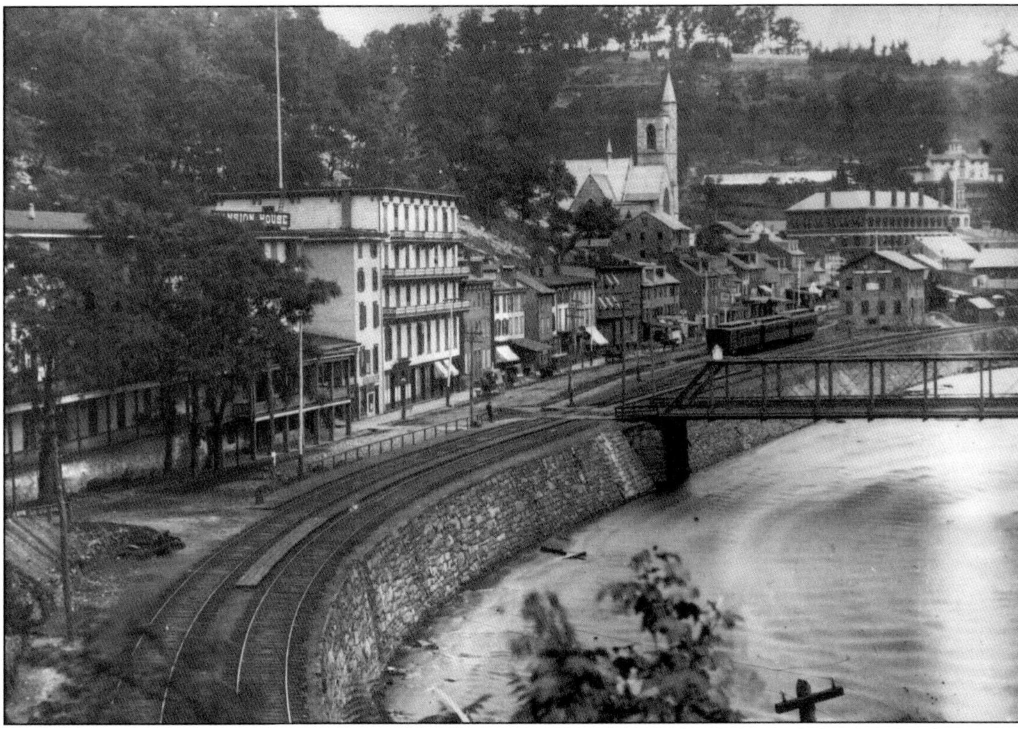

Mauch Chunk's 300-room Mansion House Hotel is shown on the left, and the footbridge across the river from the Lehigh Valley train station is on the right. (Courtesy Ray Holland.)

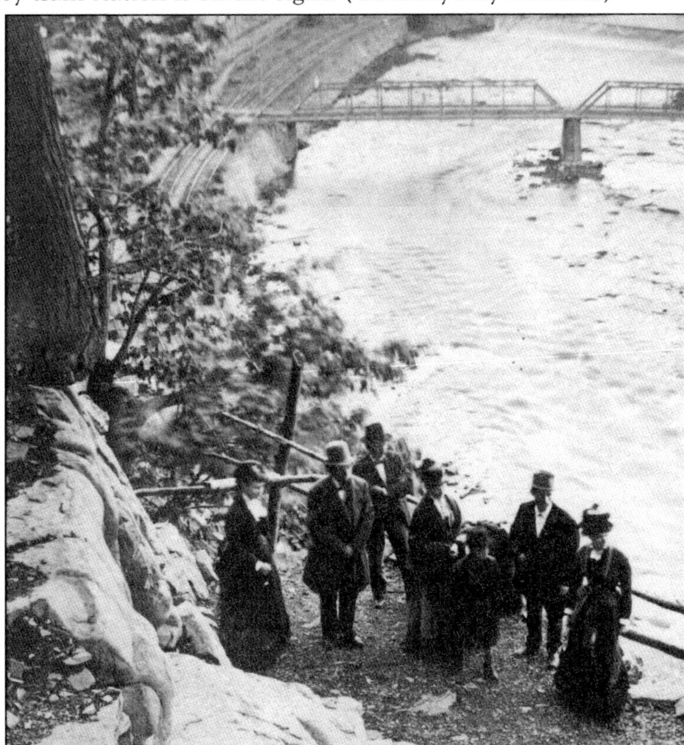

This lookout is downstream from the footbridge. Here, 19th-century tourists enjoy the splendid view along the Lehigh River. (Courtesy Ray Holland.)

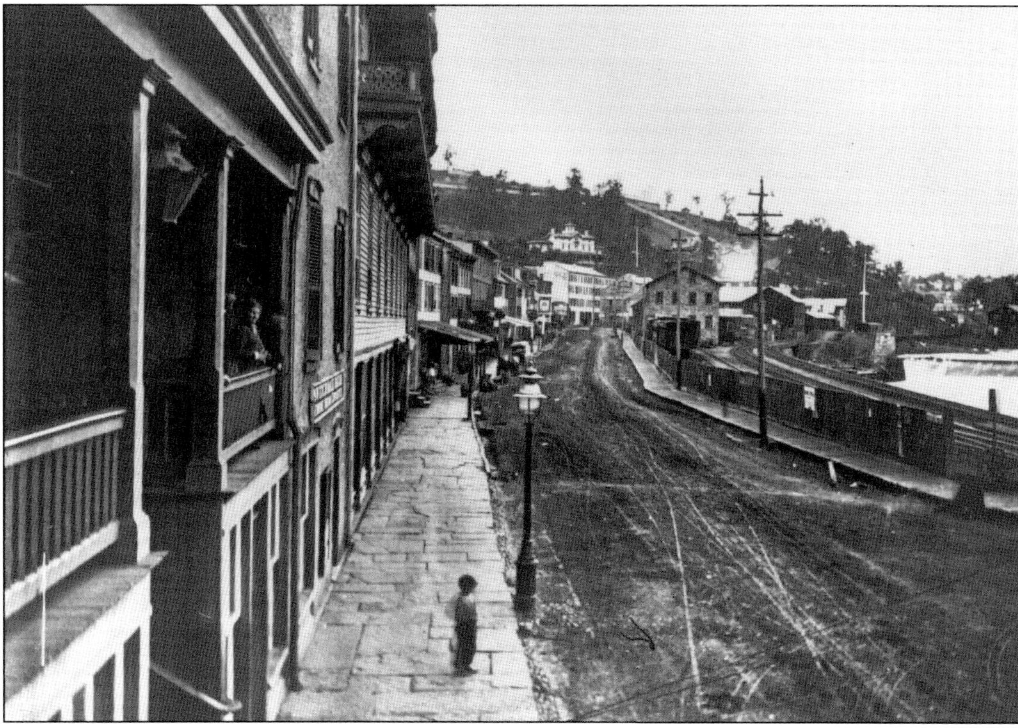

Looking toward the town center along Susquehanna Street, one can see the facade of the Mansion House Hotel on the left and the Lehigh River on the right. (Courtesy Ray Holland.)

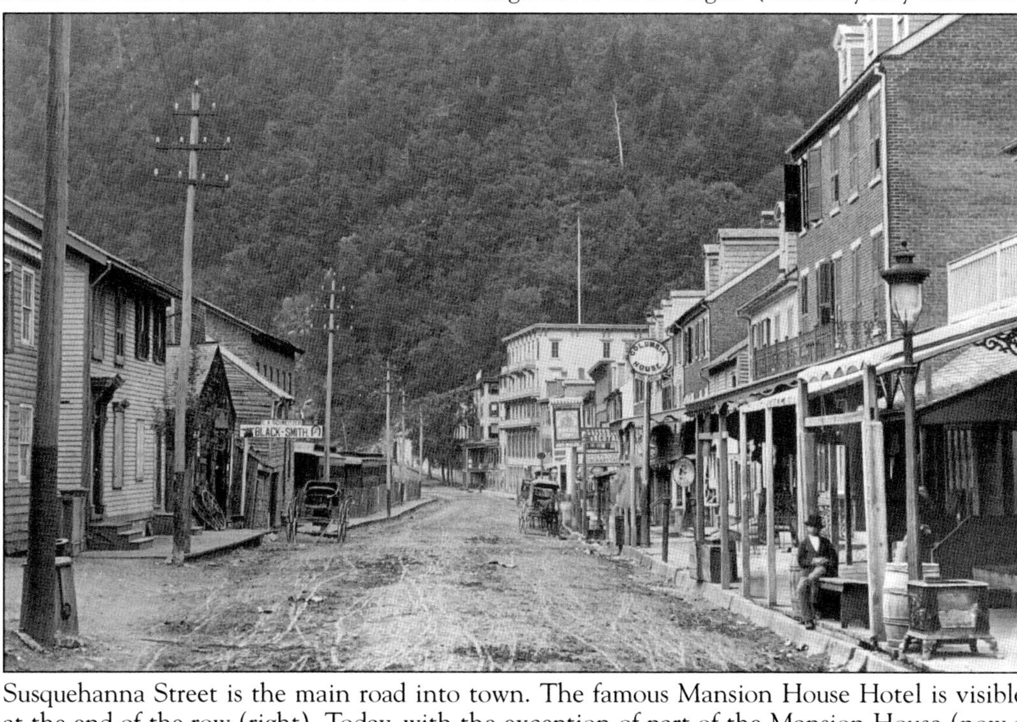

Susquehanna Street is the main road into town. The famous Mansion House Hotel is visible at the end of the row (right). Today, with the exception of part of the Mansion House (now a beer distributor), all the buildings are gone. (Courtesy Ray Holland.)

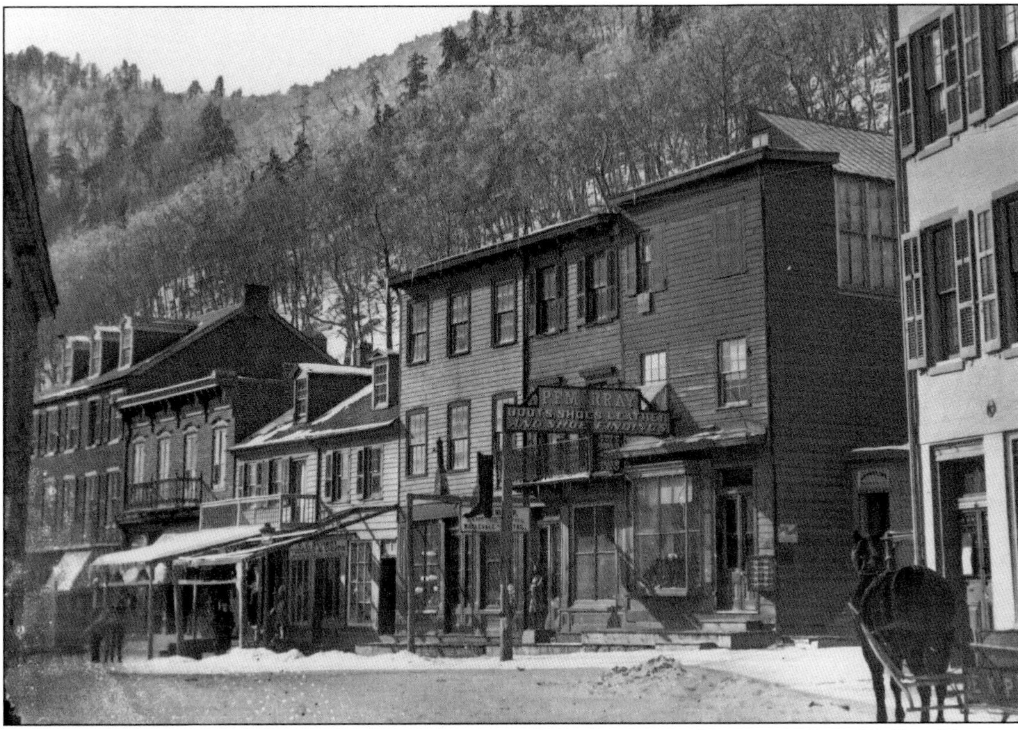

Shown are some of the many shops along Susquehanna Street. (Courtesy Ray Holland.)

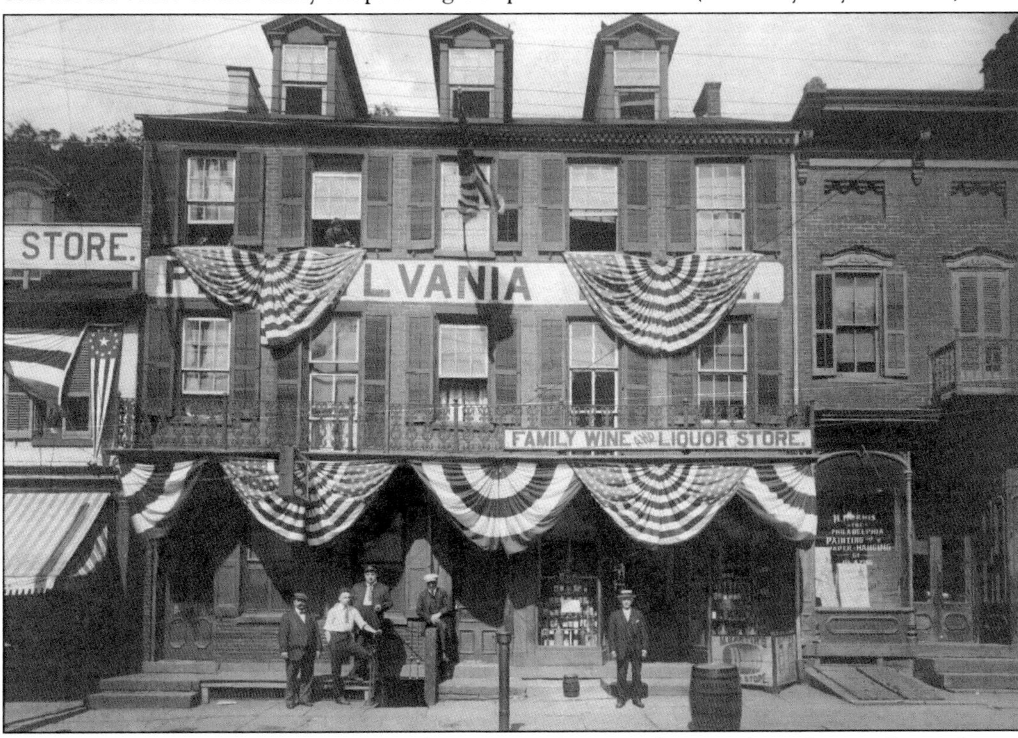

In the late 19th century, the Pennsylvania Hotel was a thriving hostelry on Susquehanna Street. (Courtesy Ray Holland.)

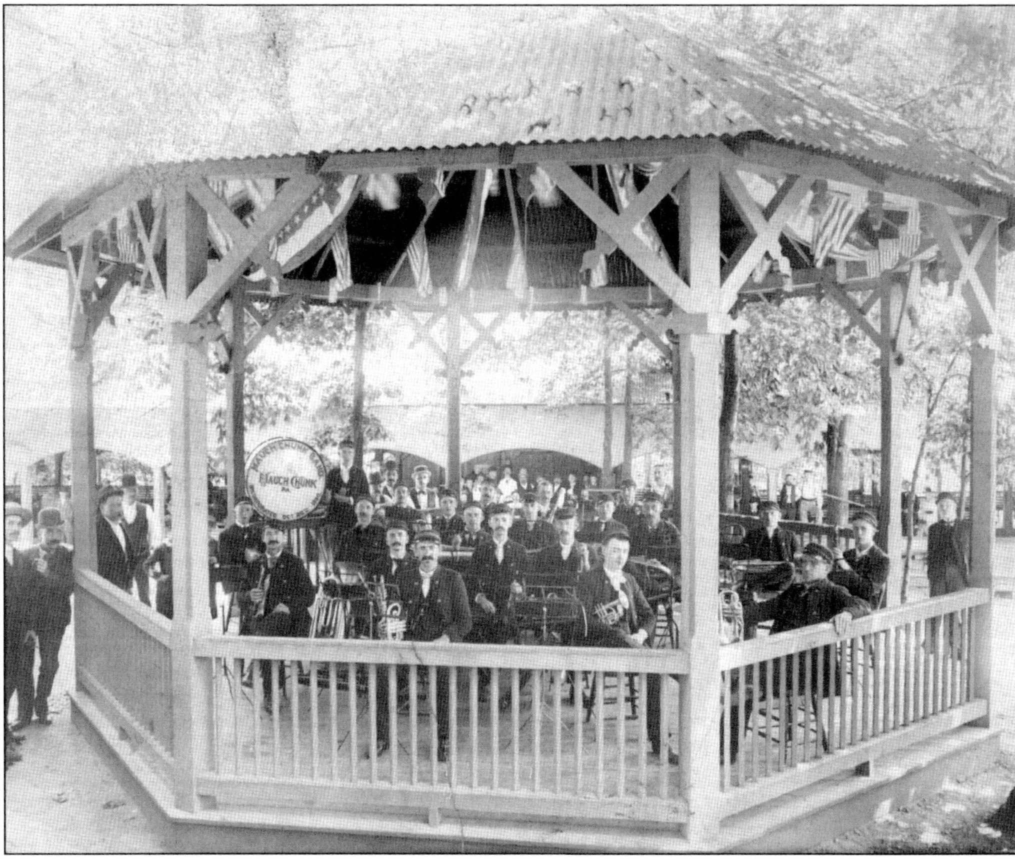

A band tunes up for a concert in the bandstand at Packer Park. (Courtesy Ray Holland.)

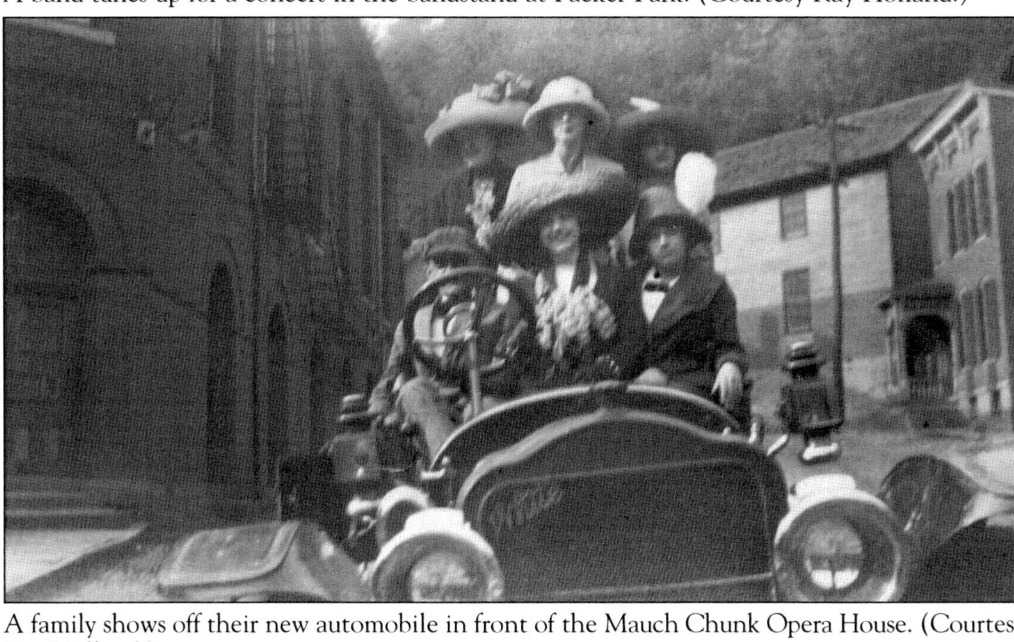

A family shows off their new automobile in front of the Mauch Chunk Opera House. (Courtesy Ray Holland.)

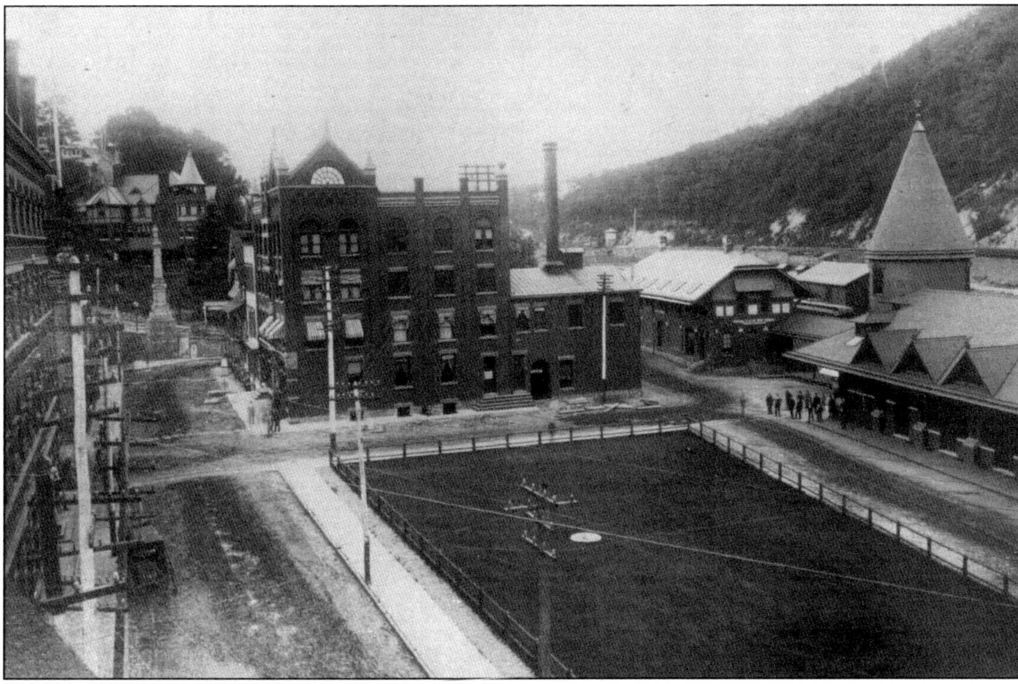

In this 19th-century view, looking along Susquehanna Street, the Lehigh Coal & Navigation Building (left) and the Jersey Central station (right) face onto a treeless Packer Park. At the top of the park is Hotel Row—the Central, Switzerland, and Armbruster Hotels. Only the Switzerland survives today.

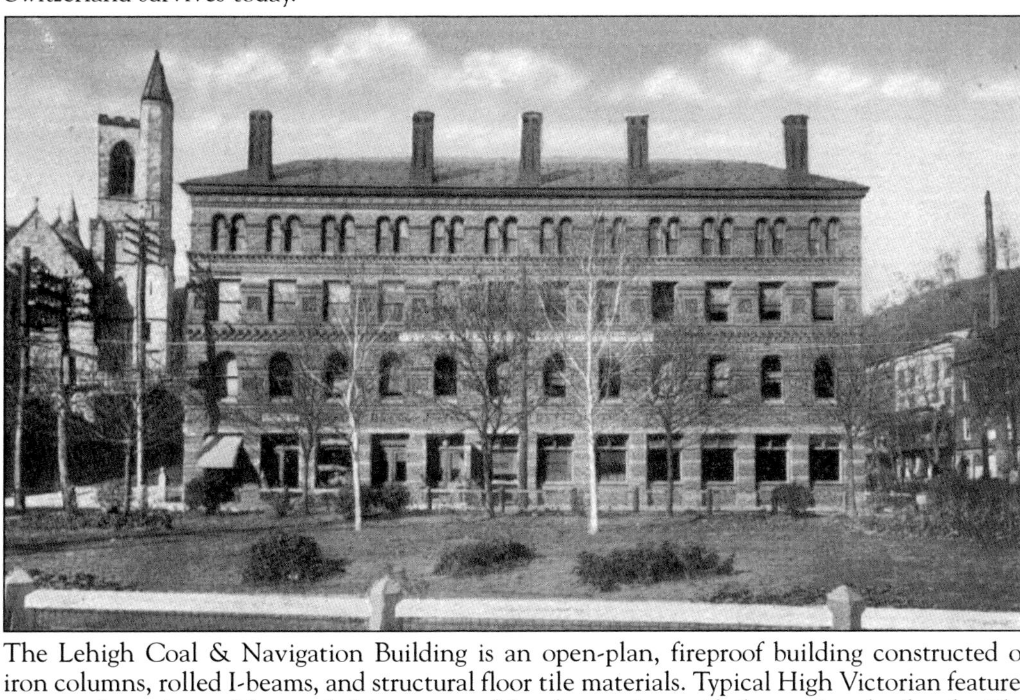

The Lehigh Coal & Navigation Building is an open-plan, fireproof building constructed of iron columns, rolled I-beams, and structural floor tile materials. Typical High Victorian features are metal cornices, elaborated chimneys, and local organic motifs decorating the facade. The architect of the 1882 building was Addison Hutton.

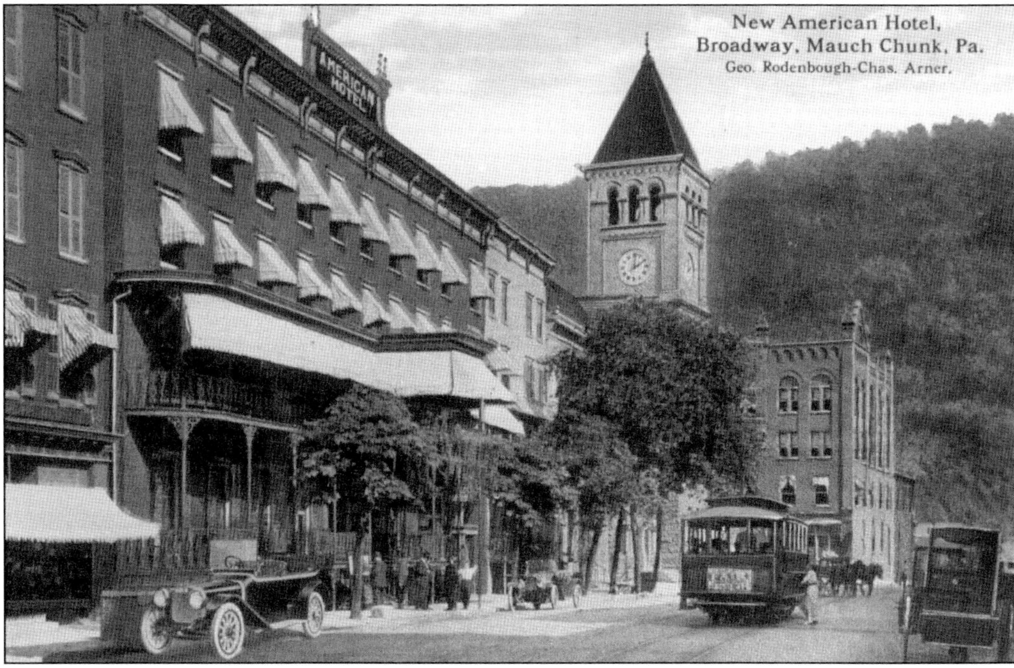

The American Hotel (left), at 20–26 Market Square, Lower Broadway, is shown in 1910. The hotel, now the award-winning Inn at Jim Thorpe, was built on the site of the 1833 White Swan following the fire that destroyed 30 buildings in the heart of Mauch Chunk on July 15, 1849. The new hotel was a brick box building with Italianate cornices, an elaborate cast-iron porte cochere, and a two-story balcony. The architect is unknown.

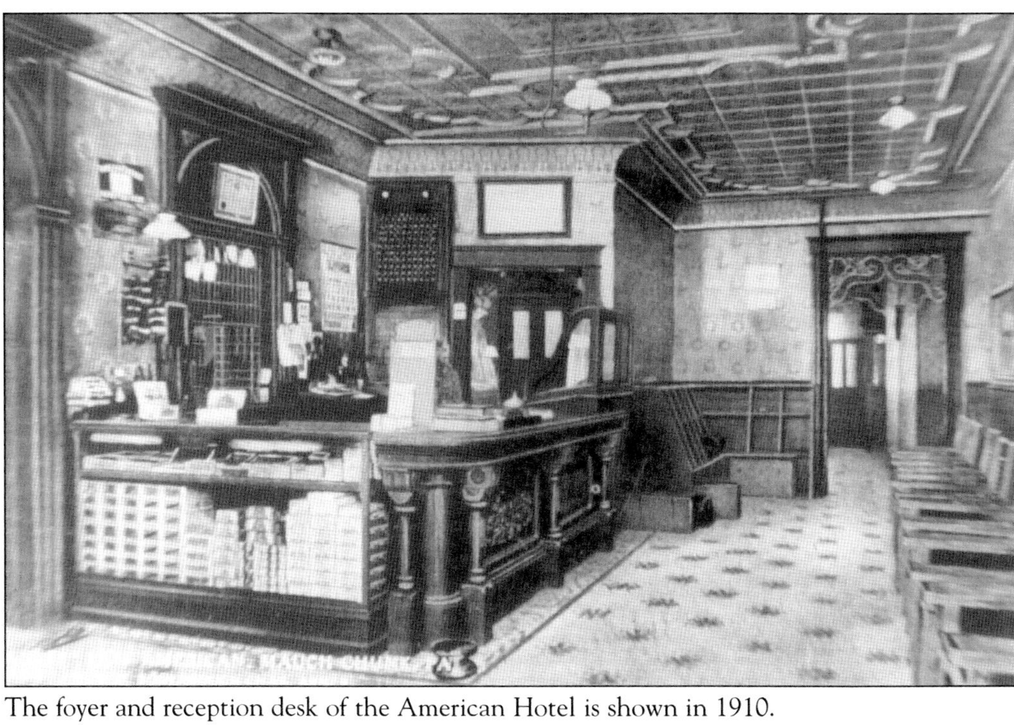

The foyer and reception desk of the American Hotel is shown in 1910.

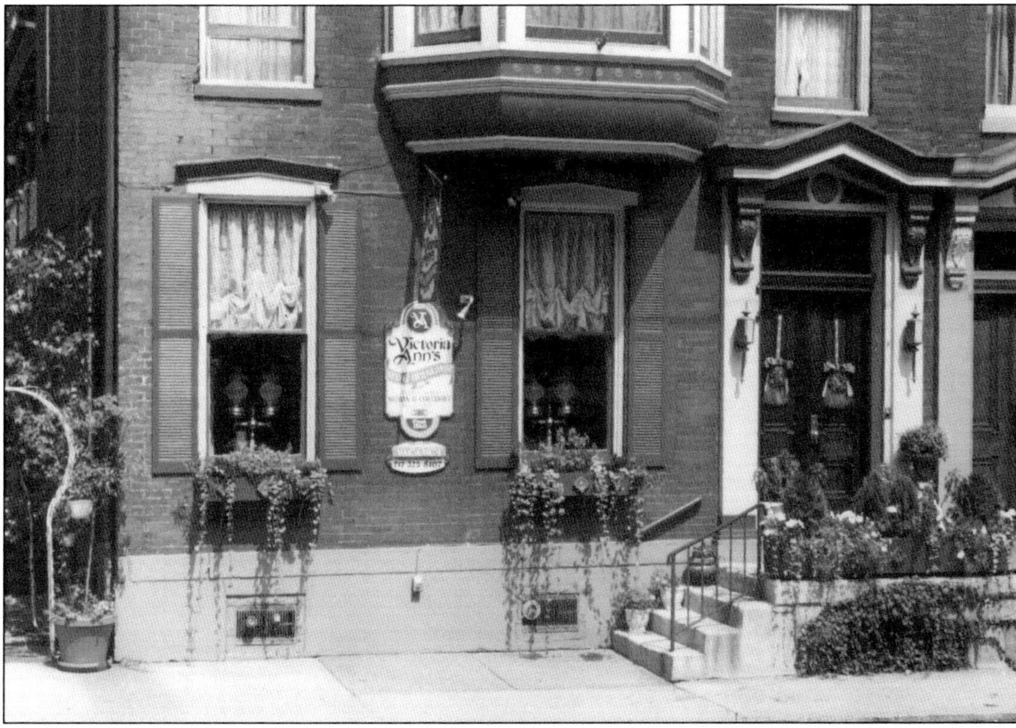

This ornate Victorian home at 68 Broadway is now Victoria Anne's Bed & Breakfast.

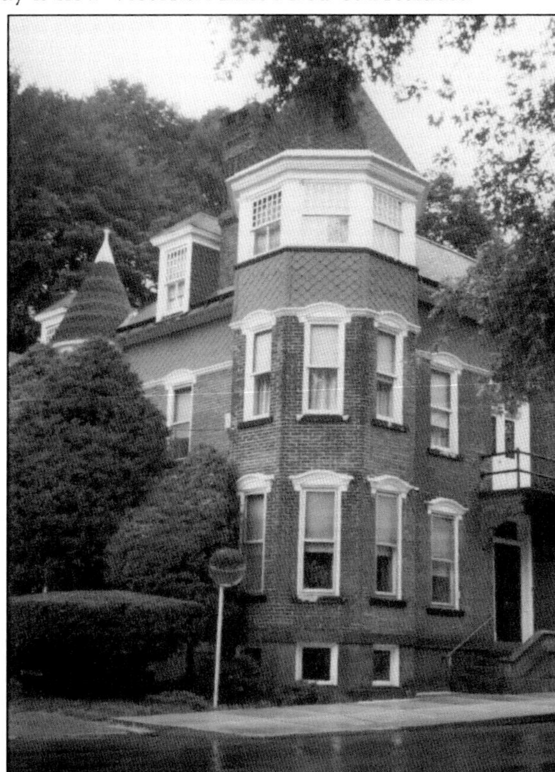

The early Queen Anne–style house at 5 West Broadway is shown with typical stylistic elements such as mixed building material textures, multiple gables, dormers, turrets, and an elaborated chimney. A recessed upper-floor porch and small square window panes are features of the upper level. Built in 1867, the building is now the Manor at Opera House Square Bed and Breakfast.

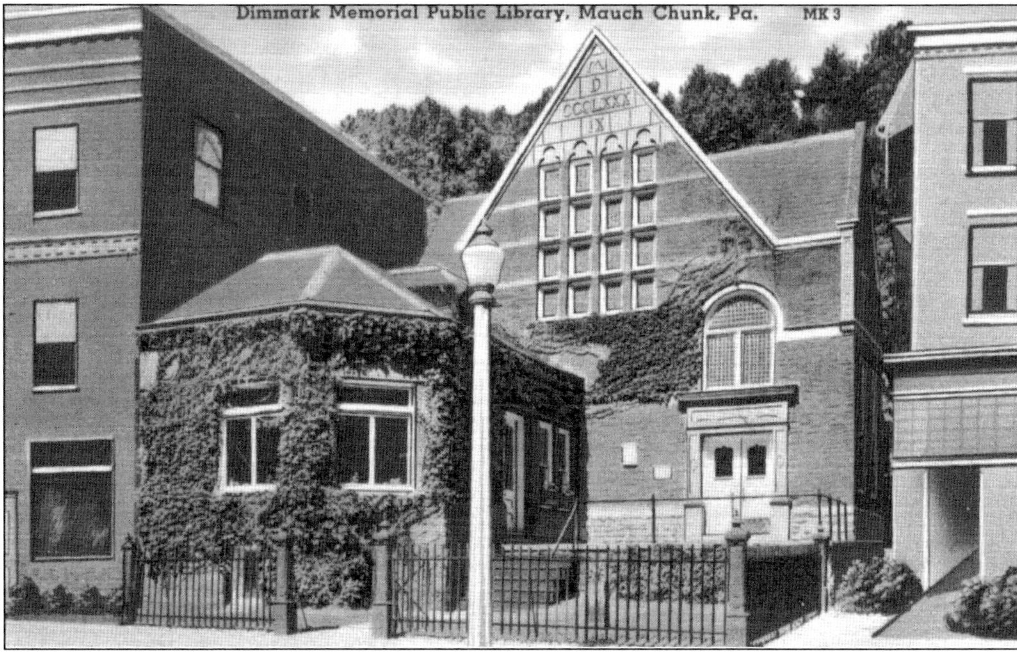

The Romanesque-style Dimmick Memorial Library was made possible by the generosity of Milton Dimmick. It was constructed in 1889 and dedicated in memory of his family.

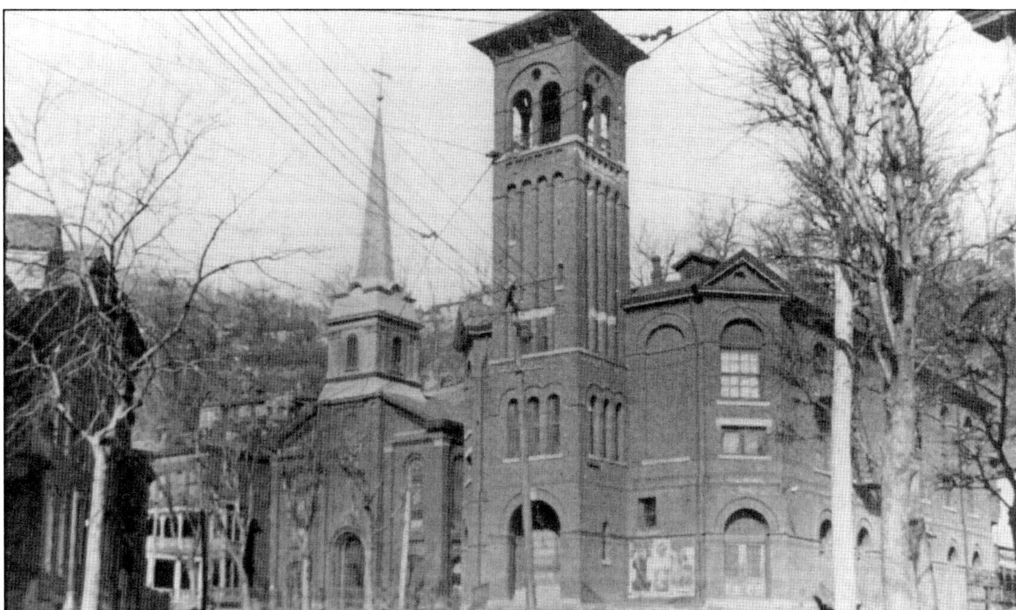

Built in 1882, the Mauch Chunk Opera House, at 14 West Broadway, is an Italian Renaissance–style building that had a central tower (now demolished). It has round-headed windows and is banded with brick-and-stone coursing. In its heyday, it was both a market and music hall. Some of the most famous entertainers of the day performed on its stage. Today, the building is home to the Mauch Chunk Historical Society. The society operates a museum and presents entertainment. The Presbyterian church (left) was dedicated on June 26, 1859. With its magnificent stained-glass windows still intact, it is currently an art museum.

St. Mark's Episcopal Church, on Race Street, was built in 1869 to replace the 1843 Norman-towered church. Designed in Gothic Revival style by Richard Upjohn with an asymmetrical tower and rose and lancet windows, St. Mark's was acclaimed "the mother of all churches in the Lehigh Valley" by a 19th-century writer. Built into Flagstaff Mountain, the church has a second-floor terrace 25 feet above Packer Square. (Courtesy Ray Holland.)

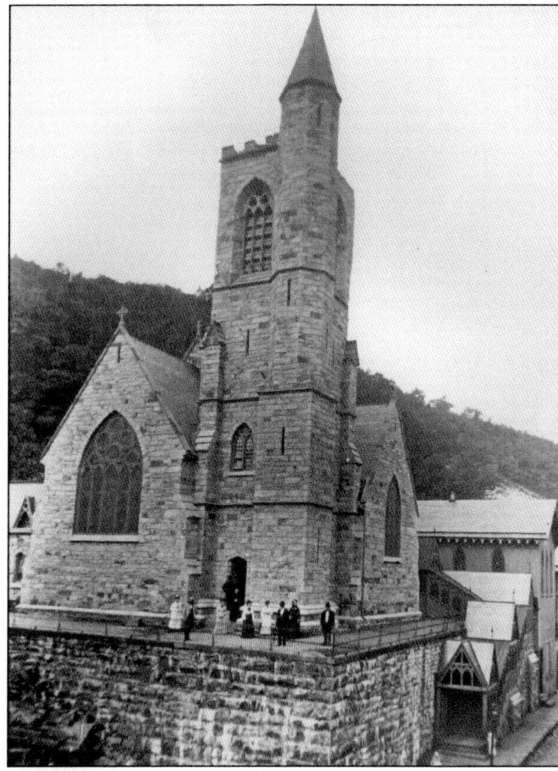

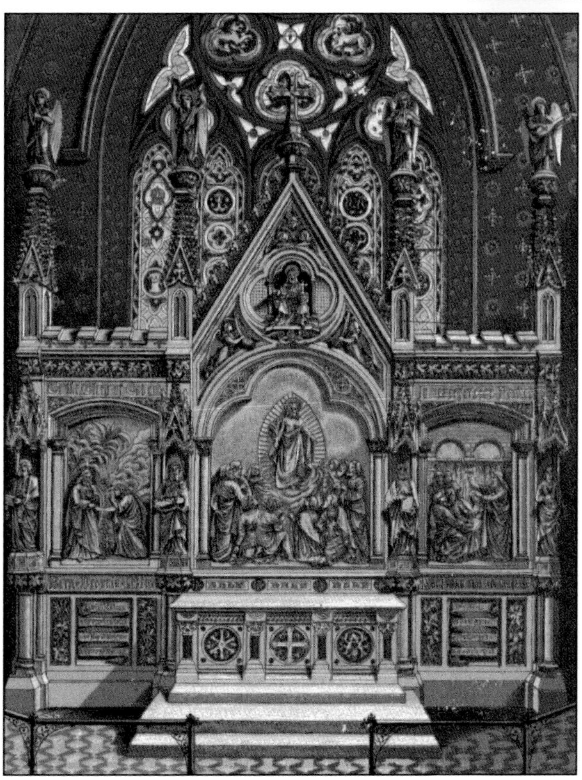

Endowed by Asa Packer, nothing was spared in the construction of St. Mark's. With its Gothic arched interior, magnificent stained-glass windows, high altar, and marble reredos, it is comparable to the finest of European churches. Guided tours are conducted of the church complex. (Courtesy Ray Holland.)

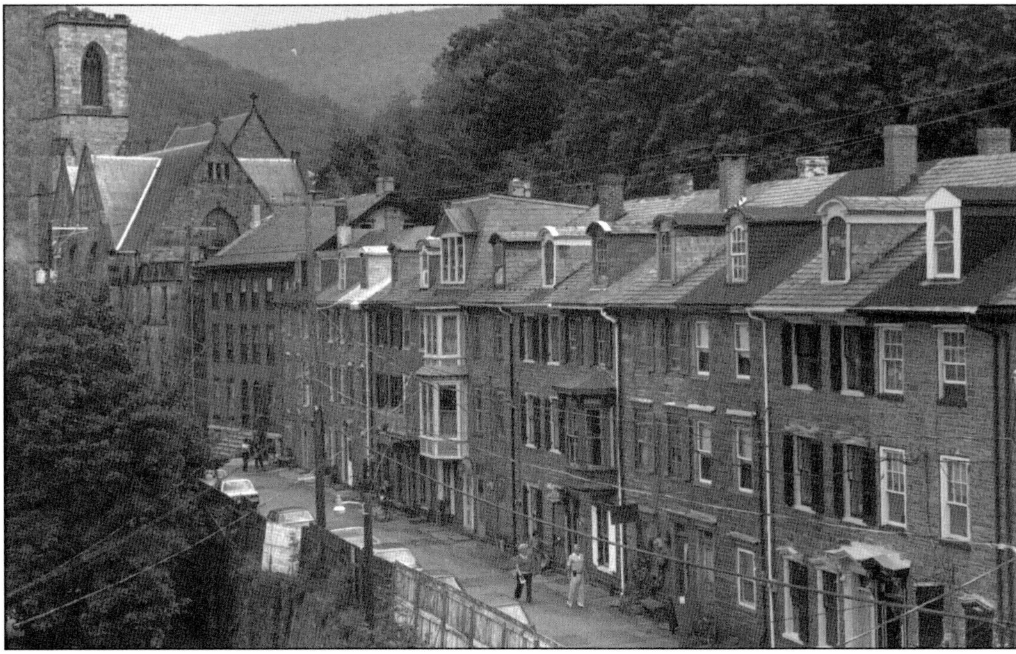

Race Street rises up from Susquehanna Street and rejoins Broadway beyond Market Square. A row of stone homes, known as Stone Row, faces the backyards of shops along Broadway. Nos. 25 to 55 Race Street were built in 1848 by Asa Packer for his workers. The houses are 14 feet wide and have gabled roofs, dormers, and varied doorways. Today, boutiques, artists, and craftspeople occupy Stone Row. St. Mark's is visible on the left. (Courtesy Agnes McCartney.)

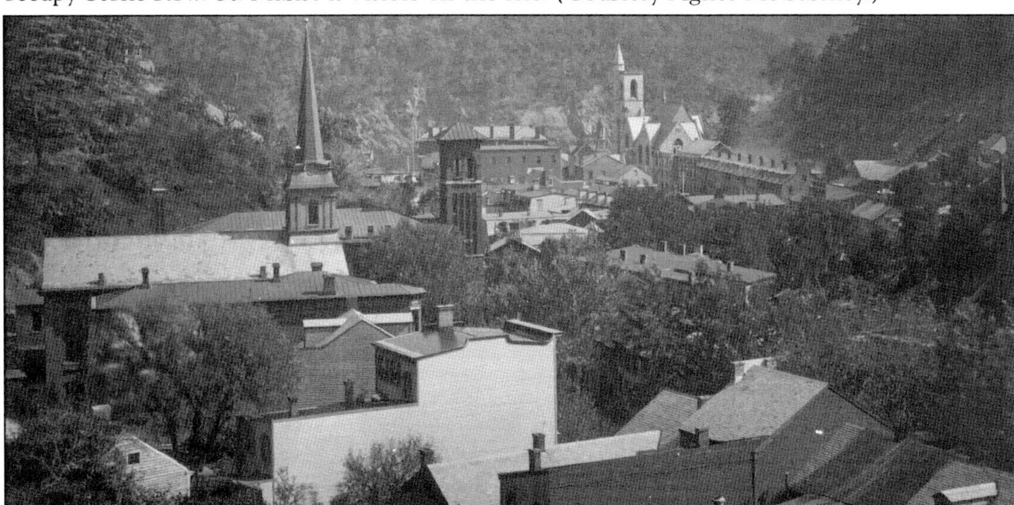

Mauch Chunk is seen from the hillside above West Broadway. If the Great Depression did one good thing for the town, it preserved its 19th-century architecture. Since the townspeople had no money to demolish or modernize landmark buildings, present-day Jim Thorpe is a Victorian time capsule. The Presbyterian church steeple (left center), the Opera House campanile (center), the elongated chimneys of the Lehigh Coal & Navigation Building (behind campanile), St. Mark's (background right), and Stone Row (right of St. Mark's) can be seen in this photograph.

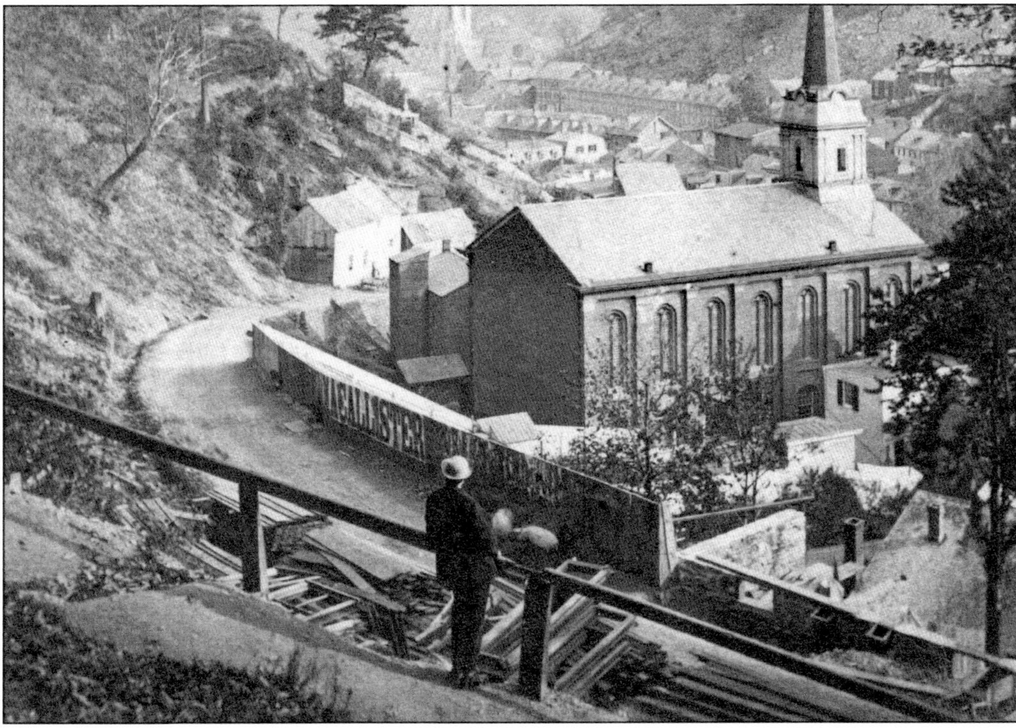

This 19th-century carriageway and terraced gardens are above the Presbyterian church. The carriageway is now known as Hill Road. (Courtesy Ray Holland.)

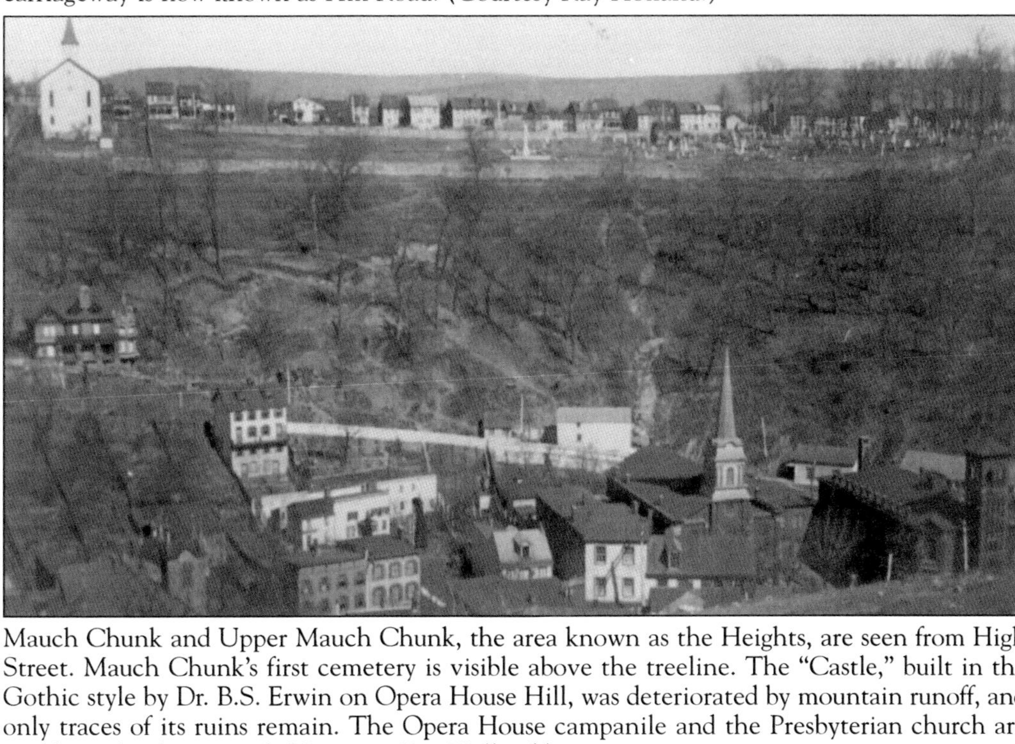

Mauch Chunk and Upper Mauch Chunk, the area known as the Heights, are seen from High Street. Mauch Chunk's first cemetery is visible above the treeline. The "Castle," built in the Gothic style by Dr. B.S. Erwin on Opera House Hill, was deteriorated by mountain runoff, and only traces of its ruins remain. The Opera House campanile and the Presbyterian church are visible in the foreground. (Courtesy Ray Holland.)

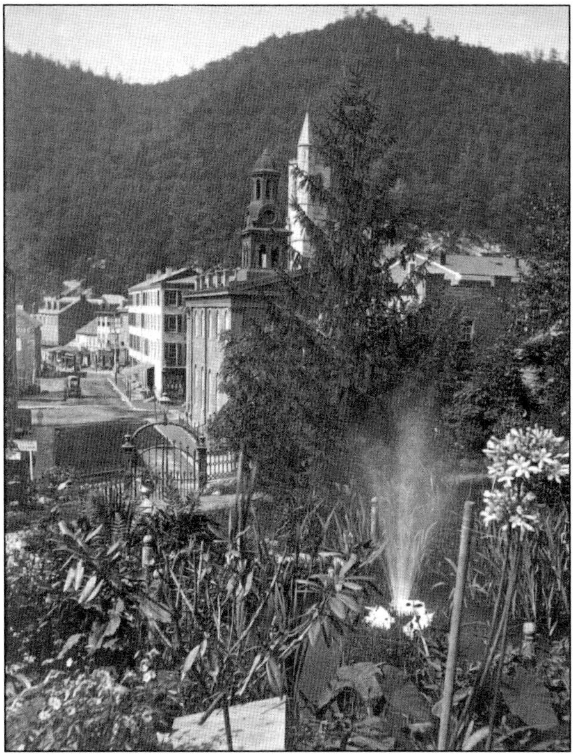

Mauch Chunk is seen from the beautiful garden of Parkhurst, formerly Whitehall, the home of Josiah White. (Courtesy Ray Holland.)

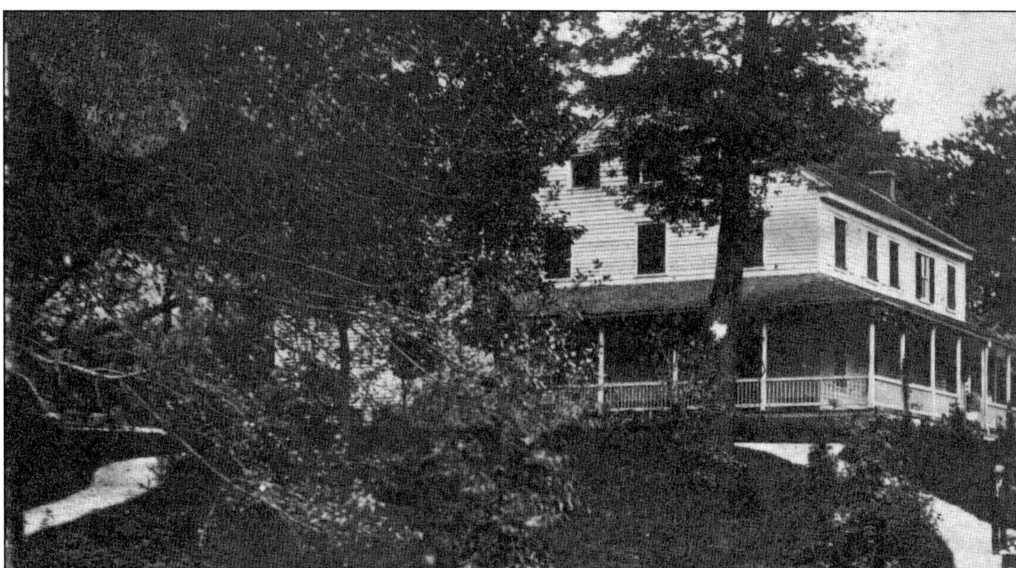

Befitting his Quaker origins, Josiah White's house was much less pretentious than the lavish mansions built later in the century by the Packers and Mahlon Kemmerer. Completed c. 1822, it was a simple white clapboard structure surrounded by an open porch. In the fashion of well-to-do Victorian England, White built a park for exotic birds and animals on the grounds. Each animal and bird had a name and was sadly mourned at its passing—even Jack Elk. When Jack attacked White, White declared, "I owe Jack no resentment. He knew no better. It was his nature." (Courtesy Ray Holland.)

John Leisenring poses outside Parkhurst. In 1861, two years after he became manager of the Lehigh Coal & Navigation Company, he occupied Whitehall. He added a third story and renamed it Parkhurst. In 1885, the house was sold to Edgar Twining, and part of it was relocated to Center Street, East Mauch Chunk. (Courtesy Ray Holland.)

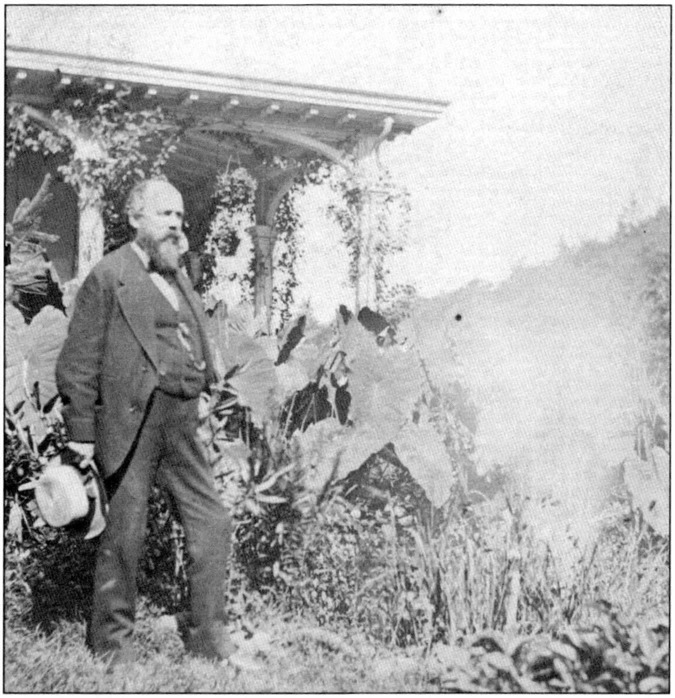

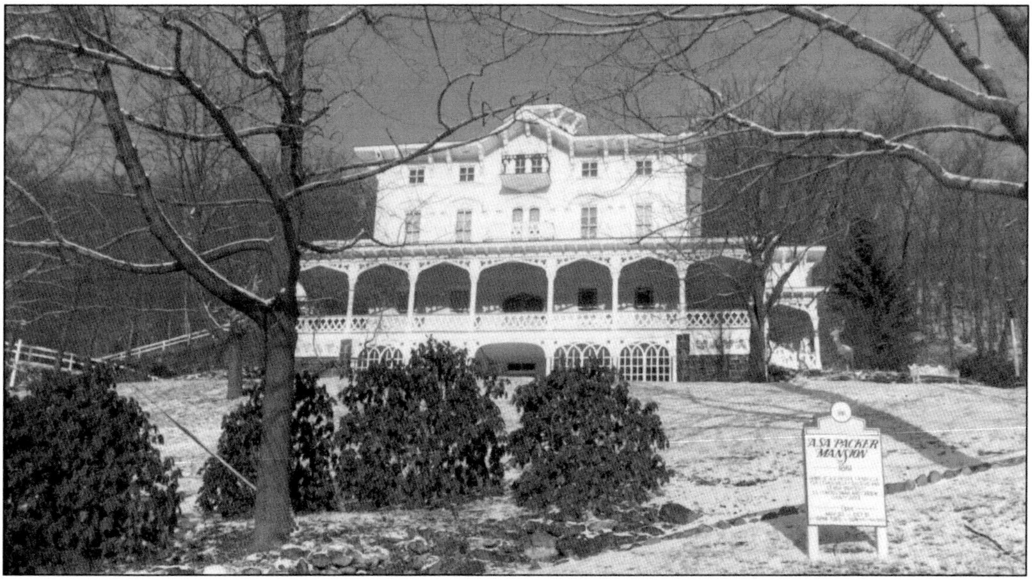

Situated on two acres of land, the Asa Packer Mansion was built in 1860 at a cost of $14,000. An 18-room, three-story symmetrical Italianate villa topped by a belvedere, it is constructed of brick with wood trim. The architect is believed to have been Samuel Sloan of Philadelphia. When Mary Packer Cummings, Asa's last surviving offspring, passed away in 1912, she left the mansion and its contents to the Borough of Mauch Chunk. It lay unused until 1953, when it was entrusted to the care of the Mauch Chunk Lions Club. The interior was cleaned, tour guides were hired, and the Asa Packer Mansion opened to the public. With its original wall coverings, hand-carved paneling, furnishings, works of art, and the personal effects of the Packers, it exists as an undisturbed microcosm of the lifestyle of the well-to-do Victorian family.

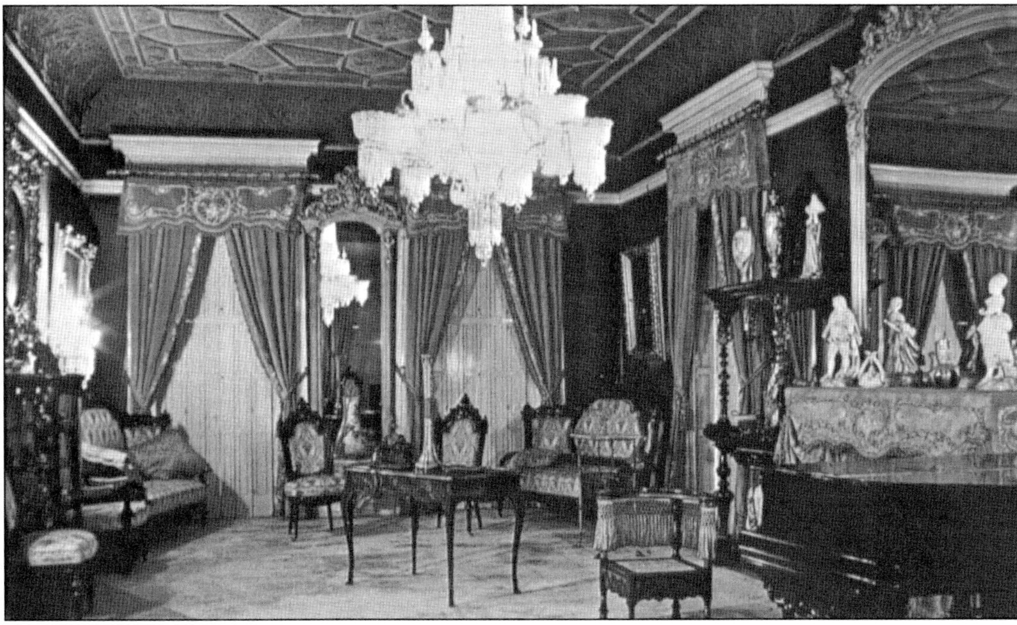

The west parlor is the biggest and most opulent room in the mansion. Large windows and mirrors, placed on all sides, give the illusion of space. The parlor was redecorated and refurbished in 1877 as part of the $85,000 renovation for the 50th wedding celebration of Asa and Sarah Packer the following year. Rosewood furniture with patterned pink silk upholstery and pink velvet draperies were purchased for the occasion. An icicle chandelier, said to have been copied for the movie *Gone with the Wind*, hangs from the ceiling with 850 pieces of crystal diffusing light in all directions. (Courtesy Asa Packer Mansion.)

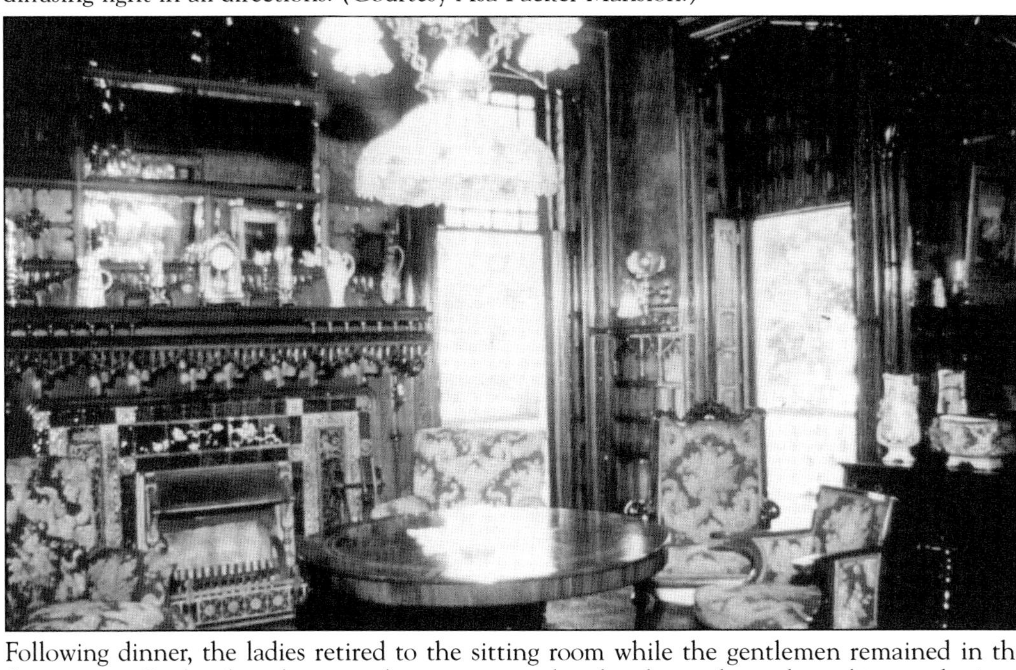

Following dinner, the ladies retired to the sitting room while the gentlemen remained in the dining room. Pocket doors between the rooms were closed so that such weighty subjects as business and politics could be discussed by the men in privacy. (Courtesy Asa Packer Mansion.)

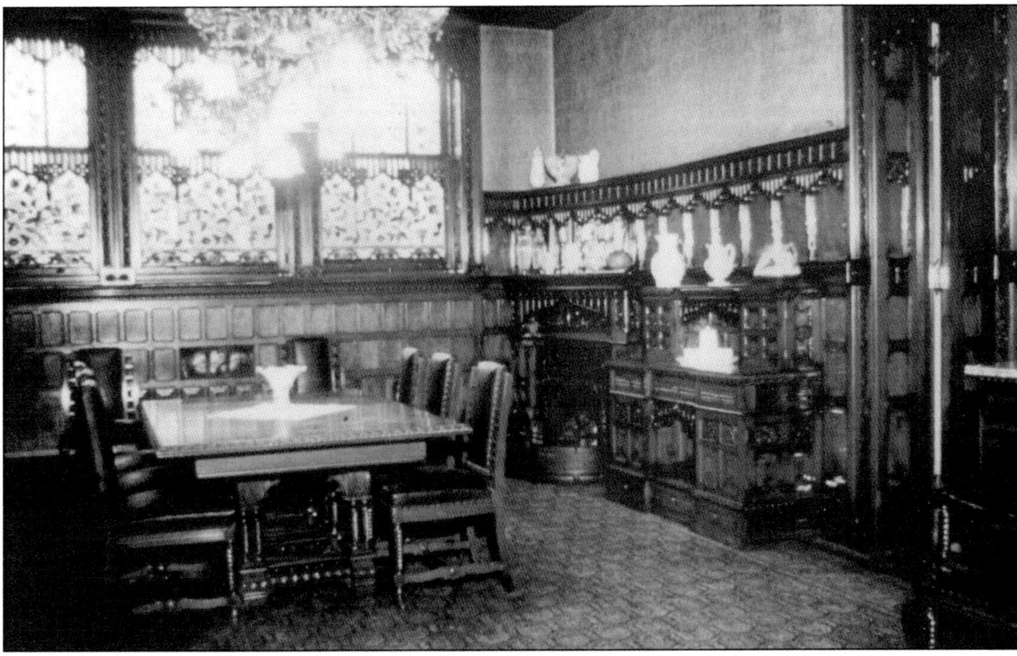

The dining room contrasts with the more feminine ladies' sitting room but, with the same Honduras mahogany paneling, is companion to it. A masculine room, it was sometimes used as a board room where Packer conducted business. (Courtesy Asa Packer Mansion.)

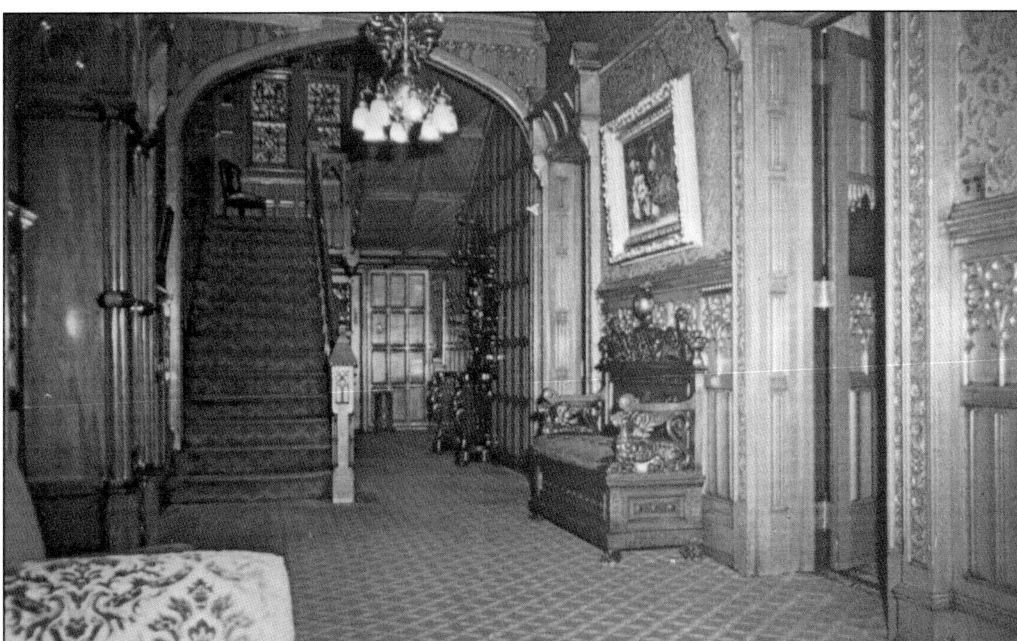

The mansion is a center-hall design with rooms called dependencies, on which the house depends for its operation. The upper walls of the hallway are stenciled gravel; the lower half is carved English post oak. Visitors arriving by horse and carriage would enter the house by the porte cochere at the rear of the residence. (Courtesy Asa Packer Mansion.)

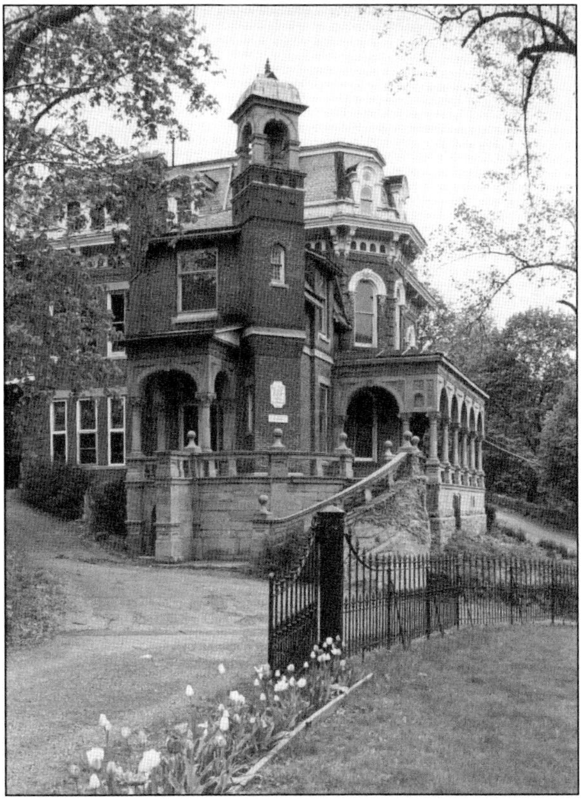

Built in 1874 by Asa Packer as a wedding present for his son Harry, the Harry Packer Mansion is a 22-room home constructed of New England sandstone and local red brick and stone with cast-iron trim. It is situated on the hillside next to the Asa Packer Mansion. With its original fireplaces, wood paneling, stained-glass windows, and Victorian furnishings, it is now an award-winning bed and breakfast.

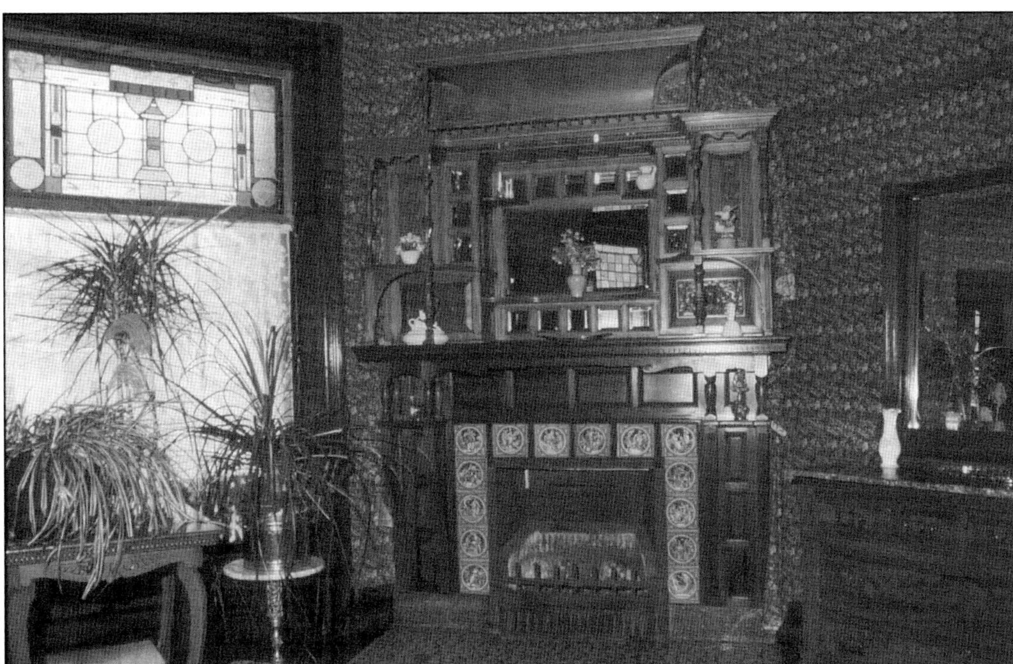

The fireplace in Harry Packer's master bedroom is decorated with tiles depicting scenes from Shakespeare's plays. The mantelpiece is carved English oak. (Courtesy Bob Handwerk.)

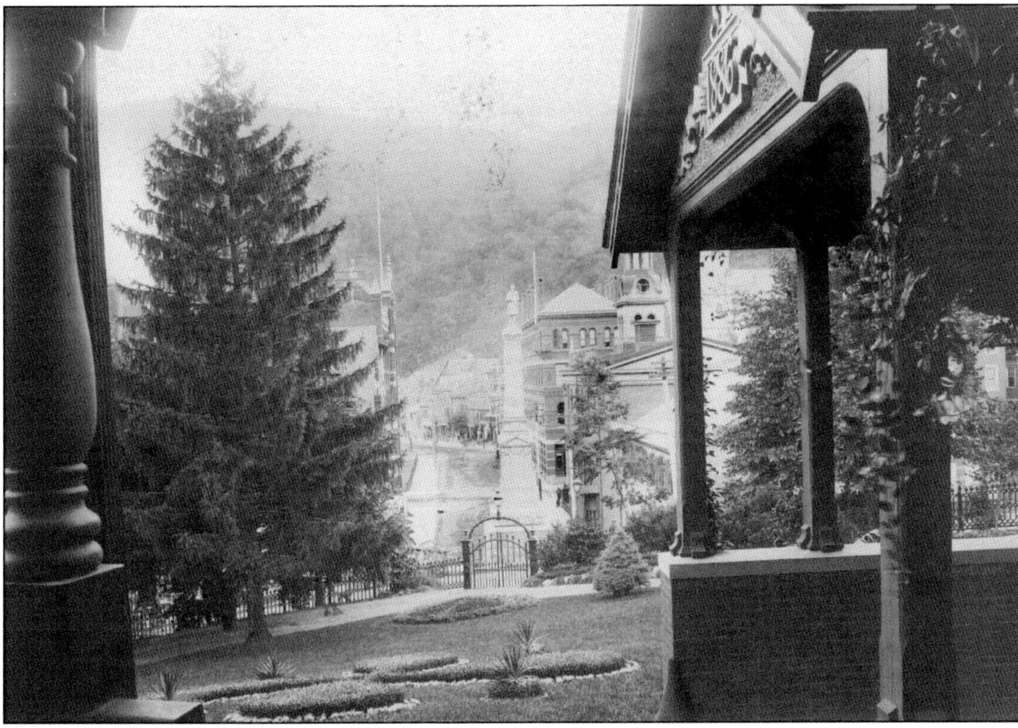

Dr. Wentz's house was built on the site of Parkhurst. It later became the Moose Hall, which was demolished in 1963. (Courtesy Ray Holland.)

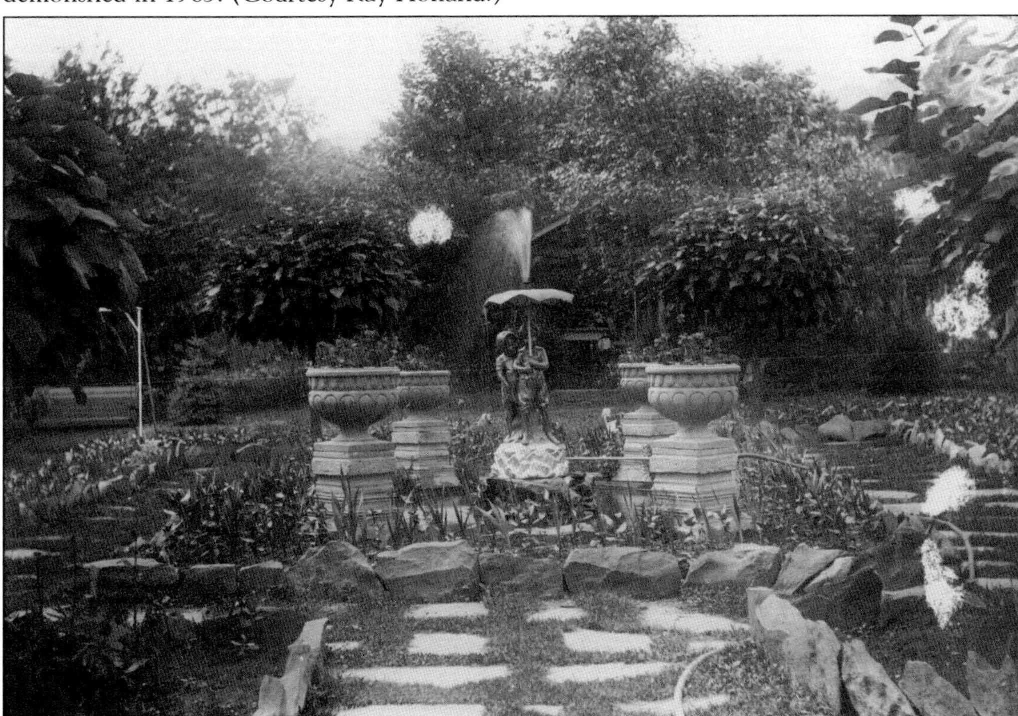

These are the lovely gardens of the estate of Mahlon Kemmerer, coal and iron entrepreneur.

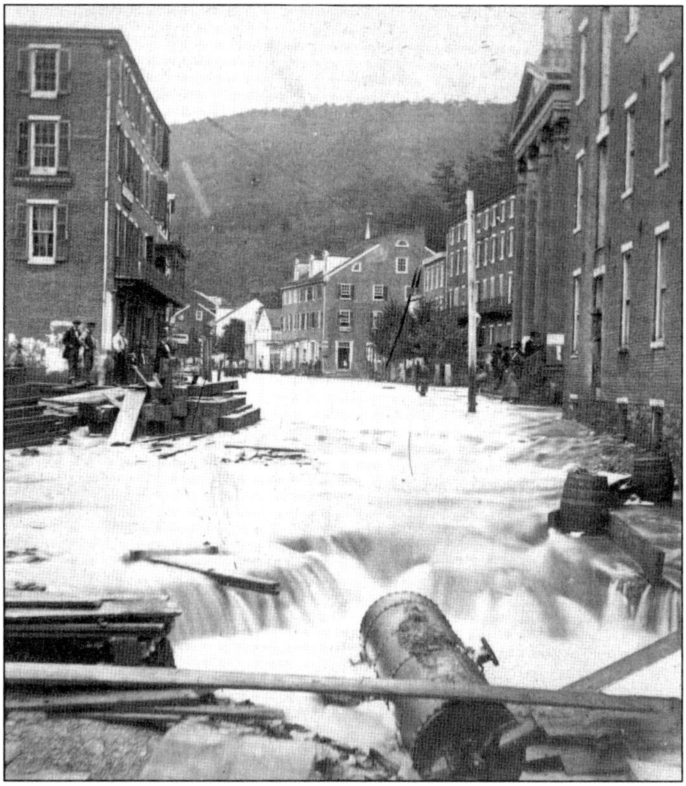

As a resident of the neighboring canal town of Weissport, George Geisel was inspired by the vagaries and power of the Lehigh River: "Yes, I am Old Man River, yet I will not tell my today's neighbor how old I am, and, if I would, who would have history to prove me? Now, in the first place, I'm a respecter of no person. I, the Old River, must take all the water that's handed to me down through the Lehigh Valley. Did you ever stop to think how many thousand miles of rain water they put on my back?"

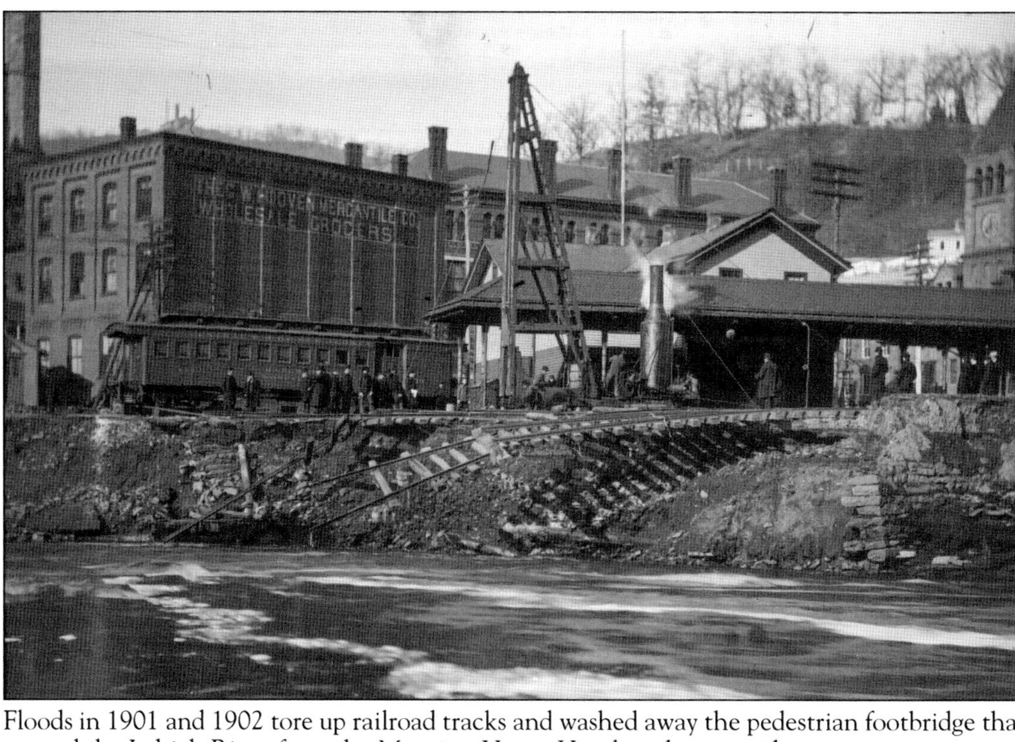

Floods in 1901 and 1902 tore up railroad tracks and washed away the pedestrian footbridge that crossed the Lehigh River from the Mansion House Hotel to the east side.

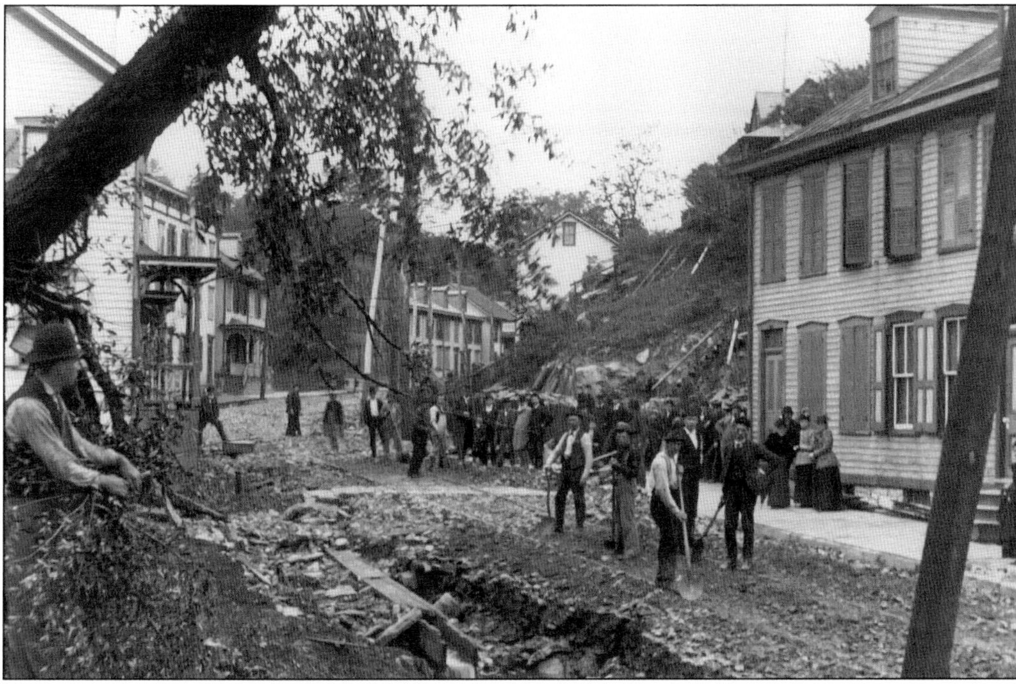

After the Mauch Chunk Creek overflowed its culvert in 1901, residents cleared debris from the trolley tracks on West Broadway. (Courtesy Ray Holland.)

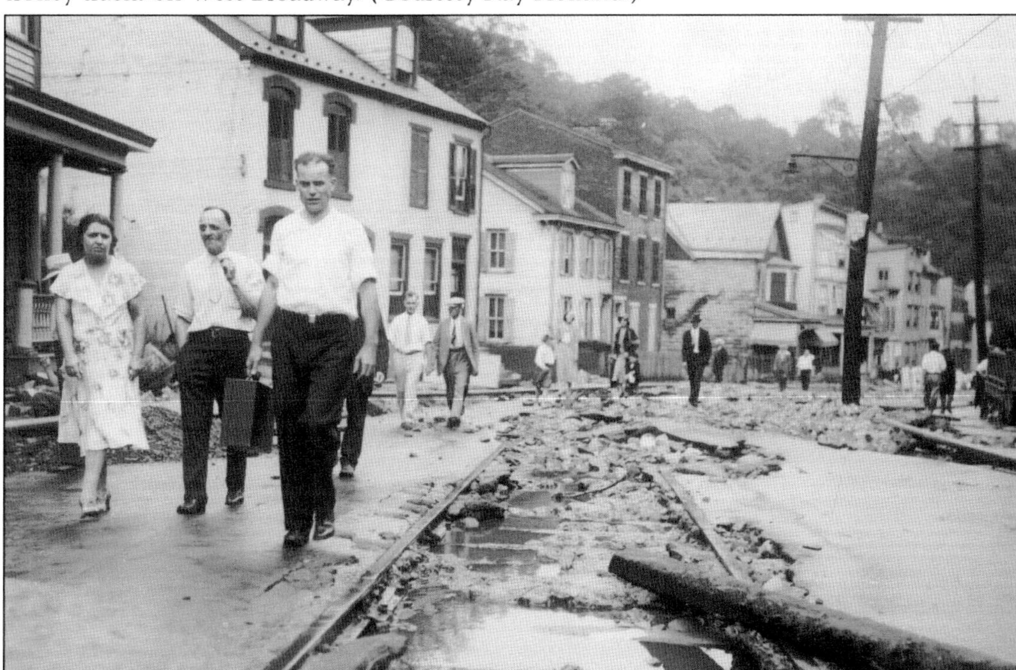

Major floods occurred in January 1841, June 1862, September 1859, October 1869, August 1901, February 1902, January 1925, June 1928, August 1933, late spring 1936, May 1942, August 1955, September 1959, August 1969, and June 1972. In 1974, flooding was controlled by damming the Mauch Chunk Creek above town.

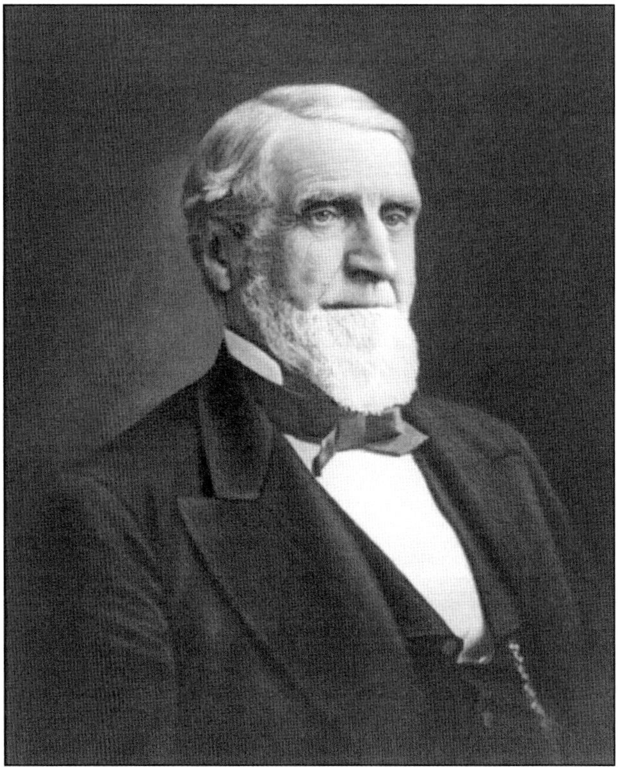

Asa Packer (1805–1879) married Sarah M. Blakslee, a farmer's daughter from Susquehanna County, in 1828. They farmed for several years before Asa set out for Mauch Chunk in 1833 to find work on the Lehigh Canal. (Courtesy Asa Packer Mansion.)

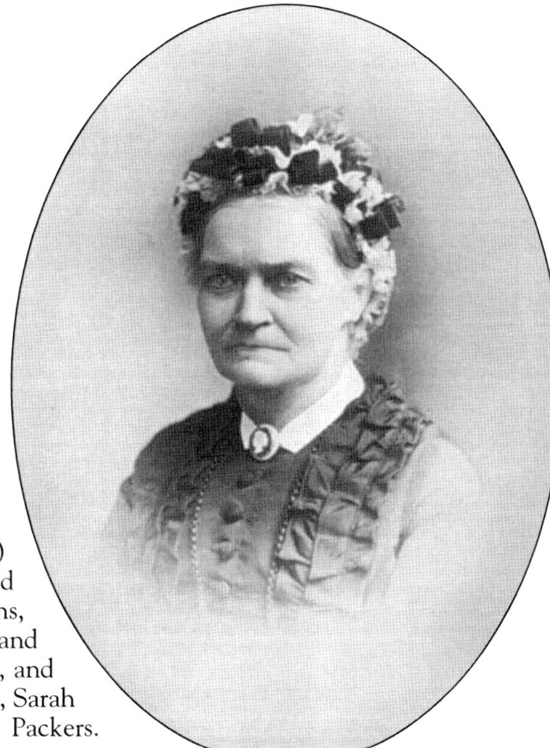

Sarah M. Blakslee Packer (1807–1882) is shown here. Sarah and Asa Packer had four surviving biological children: two sons, Robert and Harry, and two daughters, Lucy and Mary. Three daughters, Catherine, Hanna, and Gertrude, died in infancy. Marion Blakslee, Sarah Packer's half-sister, was adopted by the Packers. (Courtesy Asa Packer Mansion.)

Robert Asa Packer (1842–1883) resided in Sayre, Pennsylvania, and managed the northernmost operations of the Lehigh Valley Railroad. Today, his home is part of the Robert Packer Memorial Hospital complex. (Courtesy Asa Packer Mansion.)

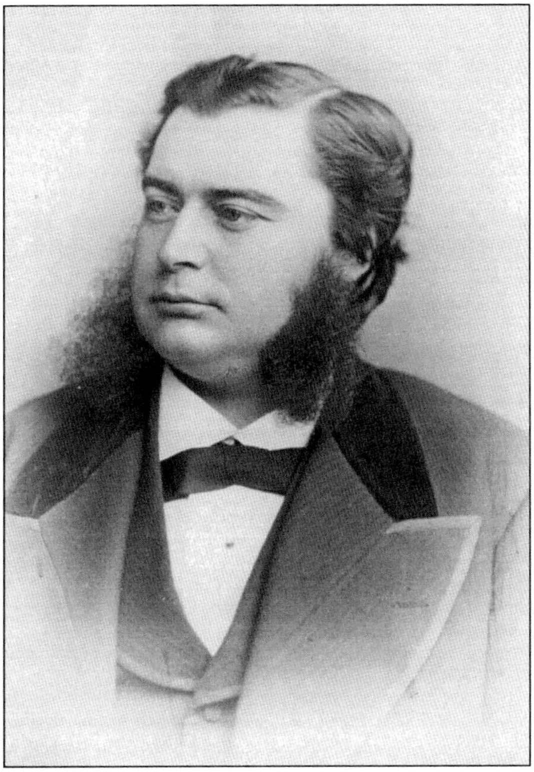

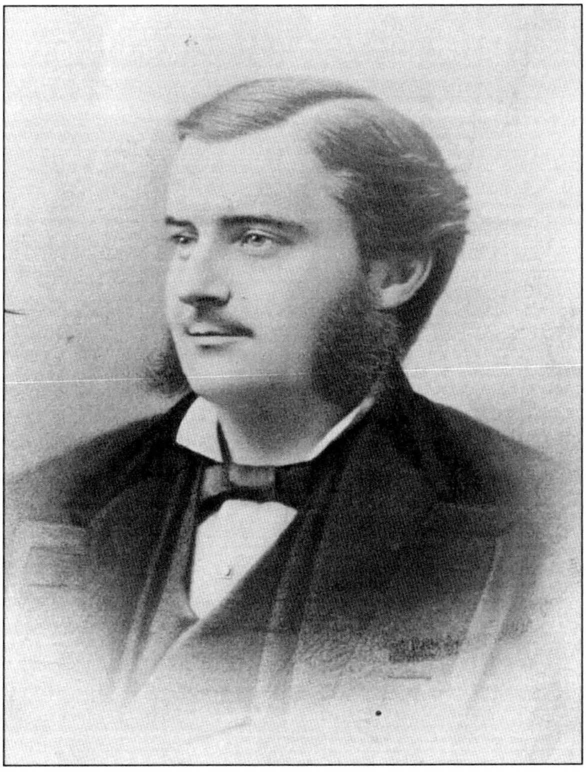

Like his brother Robert, Harry Eldred Packer (1850–1881) was involved in the operation of the railroad. When he married Mary Augusta Lockhart, his father built a mansion for the young couple, next to his own, as a wedding present (the Harry Packer Mansion). Harry died of cirrhosis of the liver at age 34. (Courtesy Asa Packer Mansion.)

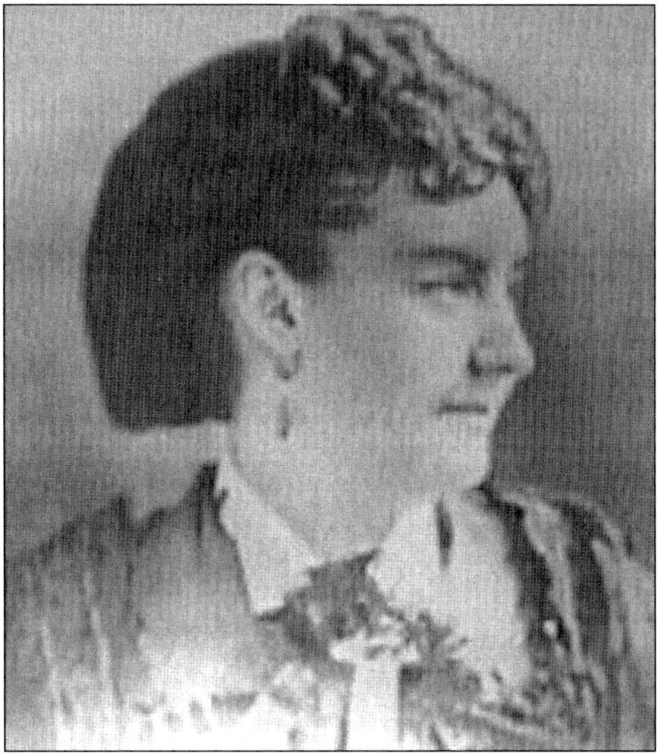

Lucy Packer (1832–1873) married Dr. Garrett Linderman and had three children: Sally, Robert, and Garrett. When Lucy preceded her parents in death, her father built the Linderman Library at Lehigh University in her memory. (Courtesy Asa Packer Mansion.)

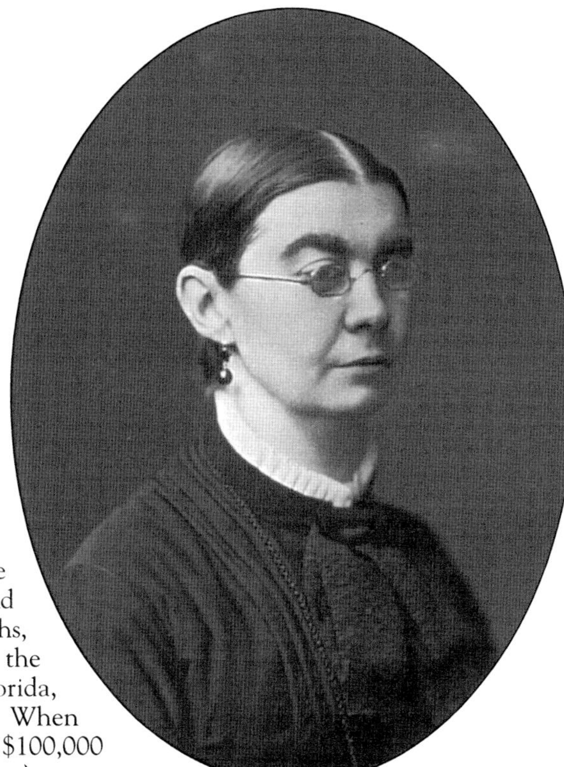

Mary Packer Cummings (1839–1912) married Charles Cummings in 1885. He was five years younger than Mary, and the marriage did not last. Within months, Charles had moved to Keystone Bluff, the Packer estate near Jacksonville, Florida, while Mary remained in Mauch Chunk. When they were divorced, Charles received a $100,000 settlement. (Courtesy Asa Packer Mansion.)

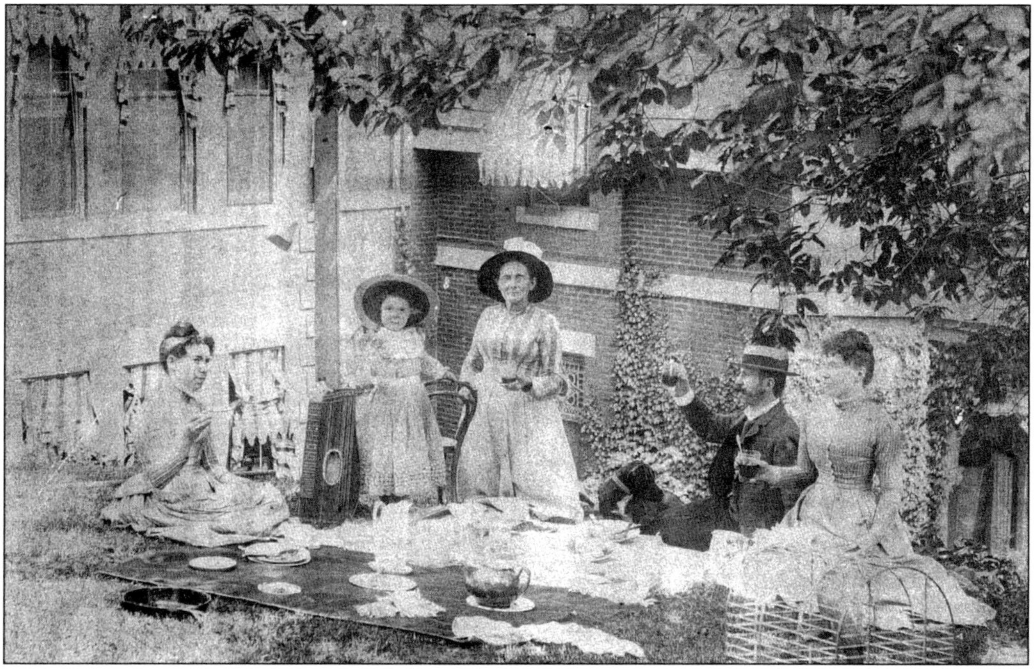

Mary Augusta Packer celebrates her adopted daughter Hazel's birthday in 1888 on the terrace behind the Harry Packer Mansion. Shown, from left to right, are unidentified; Hazel; Colleen F. Brown, governess; James Exel, butler; and Mary Augusta Lockhart Packer, Harry Packer's widow. (Courtesy Bob Handwerk.)

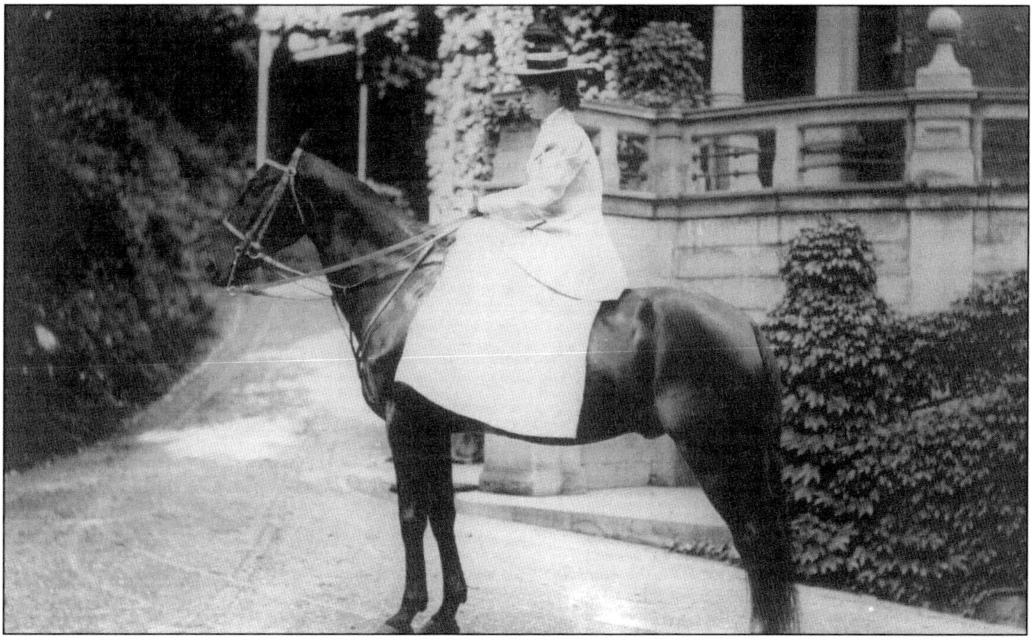

When Mary Augusta Lockhart Packer's sister Annie Hartman gave birth to an illegitimate child while staying with the young Packers, Annie was sent packing. The baby, Hazel, remained in the household and was eventually adopted by Mary Augusta. Hazel is seen on her mount outside the Harry Packer Mansion.

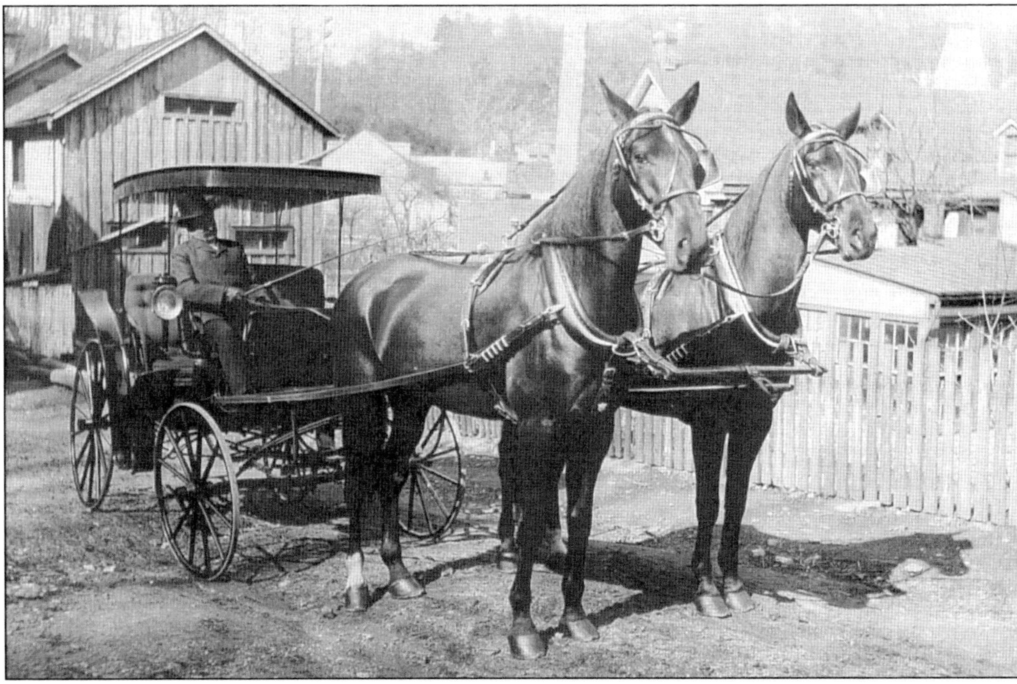

Mary Packer Cummings's carriage waits on Race Street for her to leave church at the end of the service at St. Mark's.

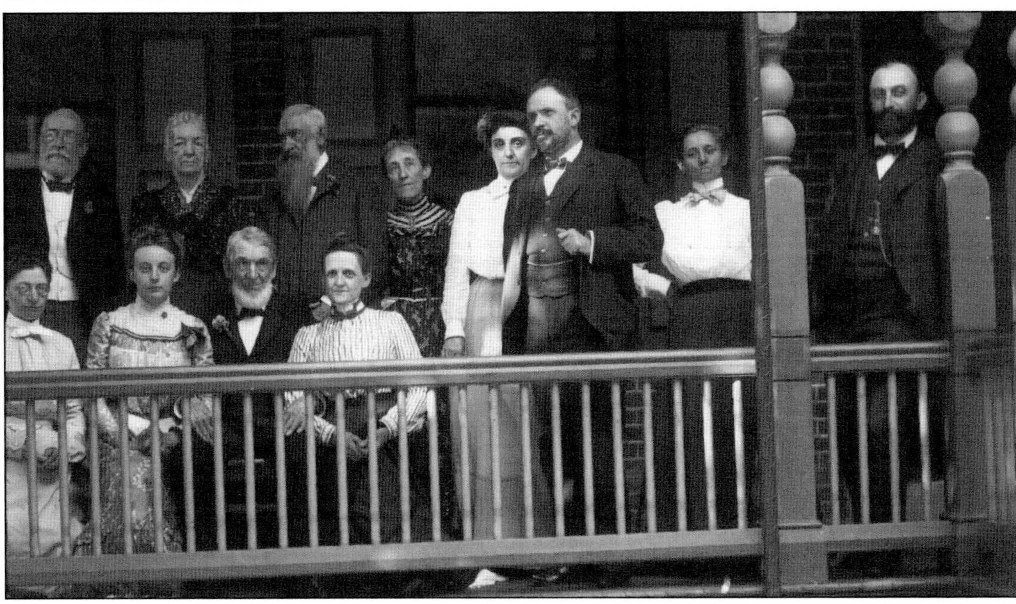

This photograph is thought to have been taken at Francis R. Sayre's 80th birthday party in 1901. Shown, from left to right, are the following: (front row) Anna Fatzinger Sayre (daughter of Francis R.), Louisa Sayre Blakslee, Francis Rodolphus Sayre, and Louisa Foster Sayre Blakslee; (back row) William. H. Sayre Jr., Anna Fatzinger Sayre (sister of William, Robert, and Francis), Robert H. Sayre, the wife of Robert H. Sayre, Connie Brown Sayre (wife of Charles Eugene), Charles Eugene Sayre, Martha Griffith (housekeeper), and Asa Packer Blakslee. The photograph was taken next to present-day Dugan's store.

"The greatest social event that has ever occurred in the Lehigh Valley, and the greatest and most important likely to occur within the present century, was the celebration by Judge and Mrs. Asa Packer of their Golden Wedding. . . . In striking contrast to the brilliant festival of Wednesday was the wedding of Asa Packer, a poor young mechanic, and Sarah M. Blakslee, the daughter of a farmer in humble circumstances, which took place 50 years ago at Dimmick Four Corners, Susquehanna County, this State." —From the *Mauch Chunk Democrat*, January 28, 1878.

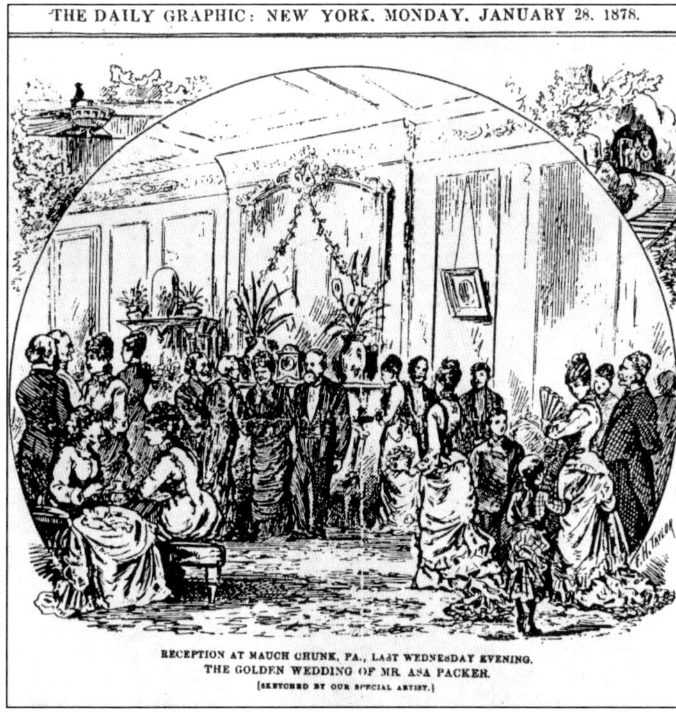
THE DAILY GRAPHIC: NEW YORK, MONDAY, JANUARY 28, 1878.

RECEPTION AT MAUCH CHUNK, PA., LAST WEDNESDAY EVENING.
THE GOLDEN WEDDING OF MR. ASA PACKER.
[SKETCHED BY OUR SPECIAL ARTIST.]

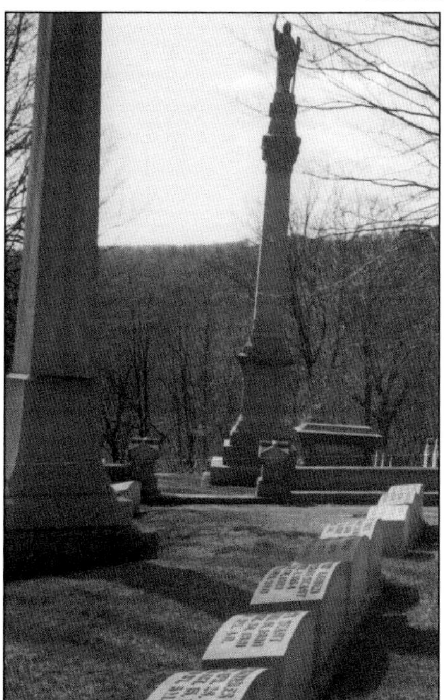

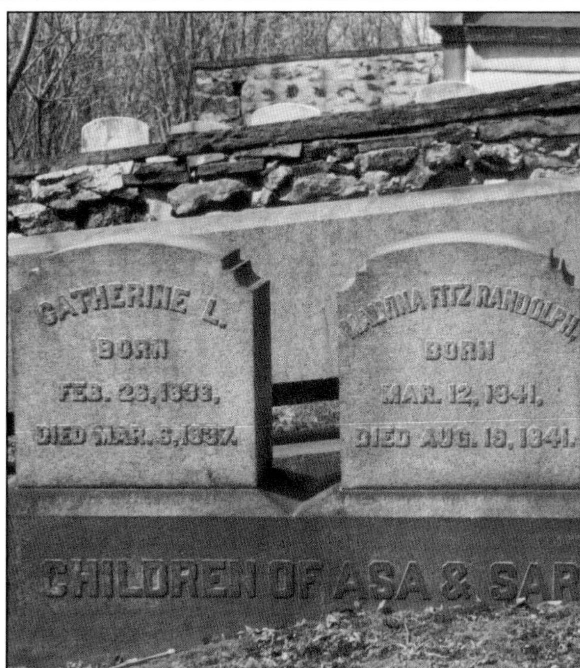

A year after the Packers' lavish golden wedding celebration, Asa Packer died. His monument, and the monuments of Mauch Chunk's early nouveau riche, may be found above their homes on Mount Pisgah and Millionaires Row in a secluded area of the old cemetery. The Packer monument is the tallest and most elaborate. Sadly, its shadow falls on three small headstones: those of Sarah and Asa's daughters, Catherine, Hanna, and Gertrude, who died in infancy.

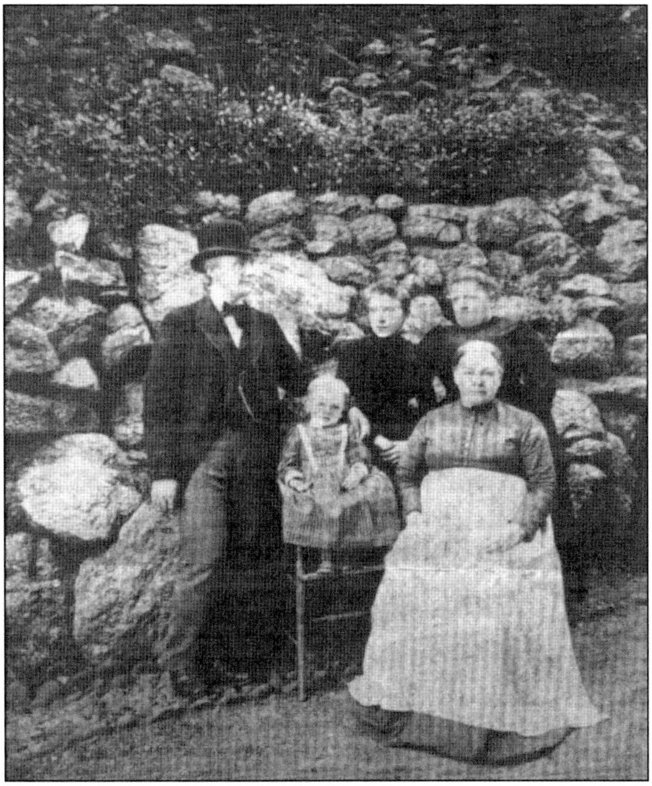

The household staff of 11 consisted of a butler, two cooks, assorted maids, drivers, gardeners, clerks and Asa's male secretary. Only three of the staff were resident: the butler, cook, and assistant cook. To make sure that "ne'er the twain shall meet," the Packers hired Irish Catholic women and German Lutheran men. The members of the staff photographed are, from left to right, James Zellner, butler; Bridget Breslin, cook; Charles Pearsall, son of Mrs. Packer's secretary David Pearsall; an unidentified maid; and Kitty Breslin, the mother of the cook.

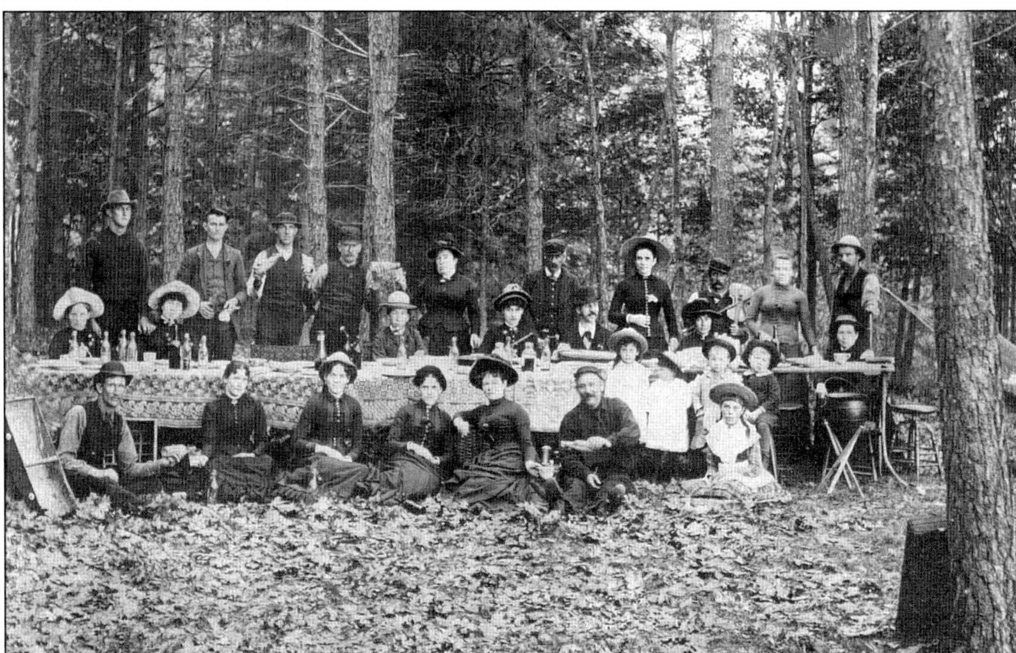

This 1890 photograph shows the hired help from Millionaires Row picnic at the Pines, near Mauch Chunk Creek. Harry Packer's butler, James Exel, played his zither at the event. (Courtesy Keith Bellhorn.)

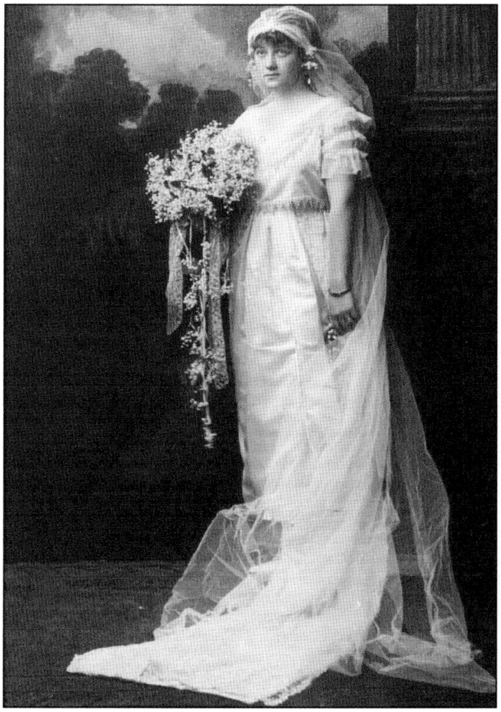

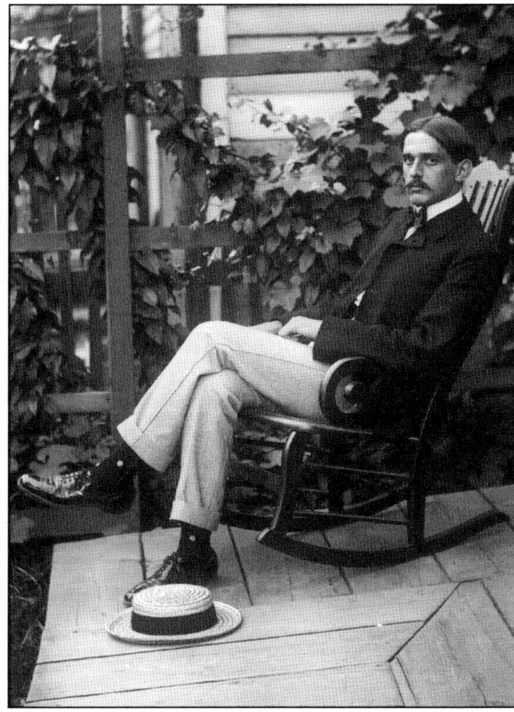

Little did this handsome family dream that their images would appear in the pages of a book. For more than a century, a box of glass slides lay gathering dust under the eaves of an old house. Discovered when the house was being renovated, the slides were offered to the Mauch Chunk Museum and give a glimpse into the lives of the subjects. Above, the bride and groom pose on their wedding day. Below are the proud parents.

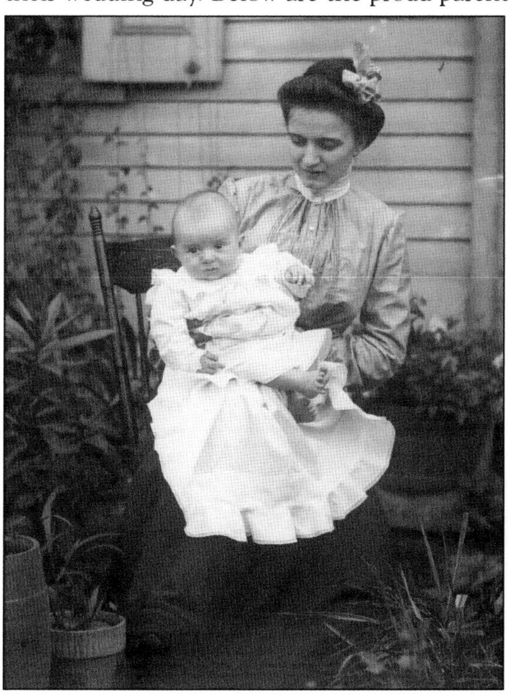

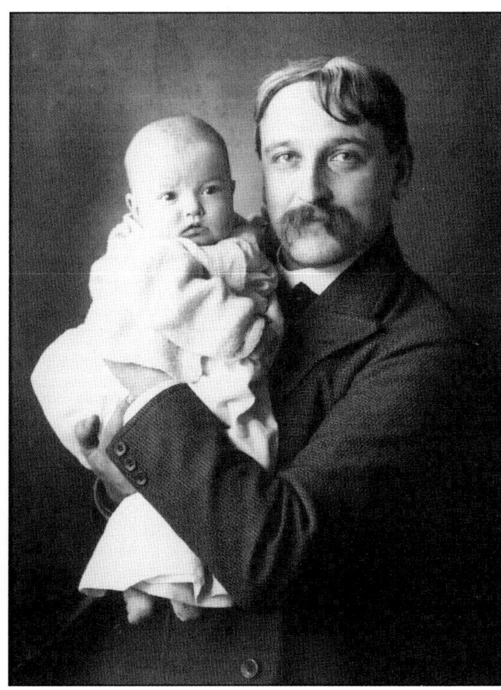

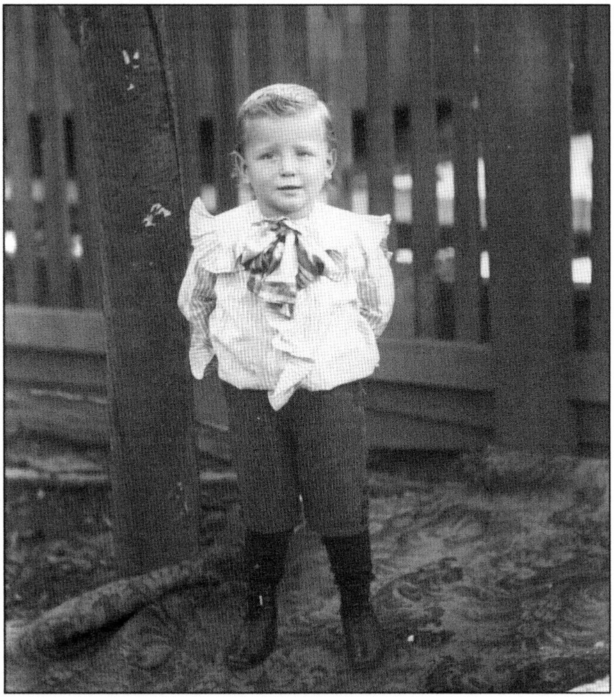

When viewing the photographs of this unidentified family, the inclination is to play detective. Is this smiling boy the same young person in the coffin below?

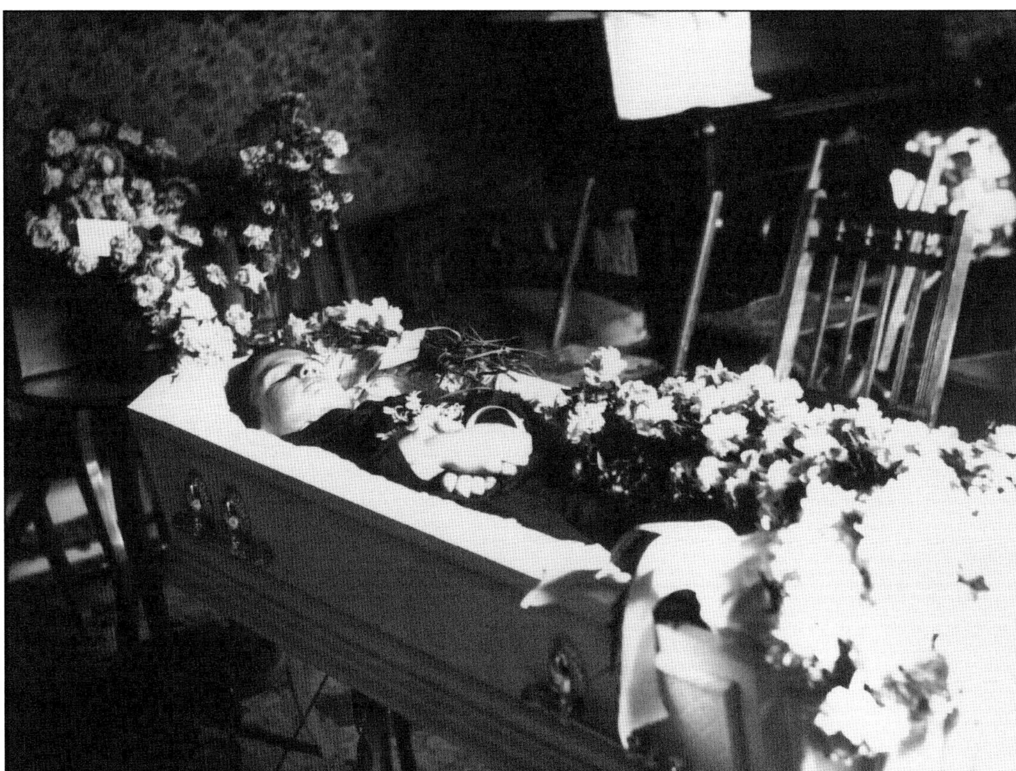

In the days before funeral parlors, the deceased was laid out in the front parlor for family and friends to pay their last respects.

A day hunting in the woods was an enjoyable adventure and provided meat for the table.

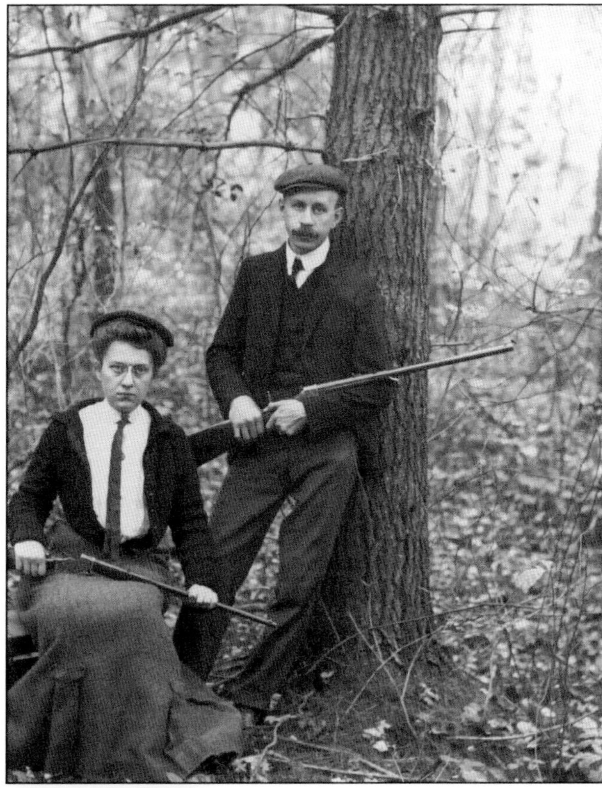

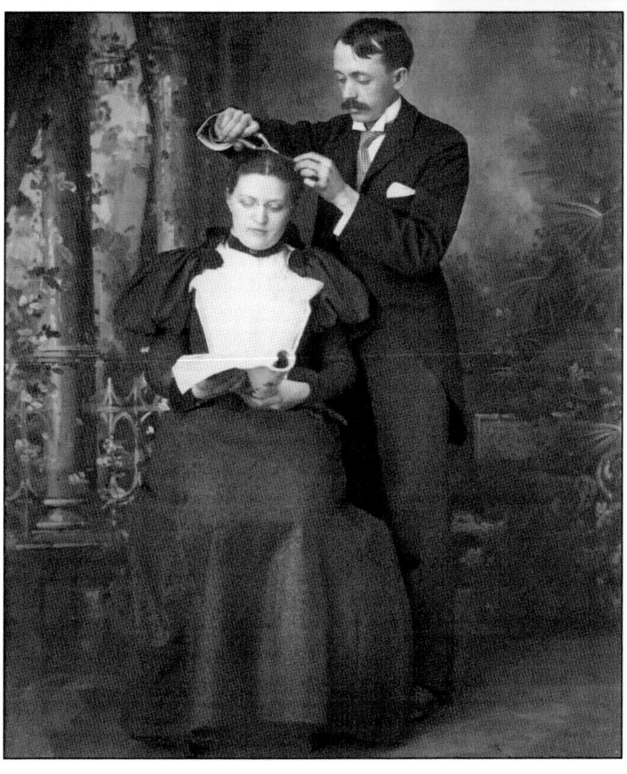

In preparation for a gala evening, a husband puts the finishing touch on his wife's hair with a curling iron.

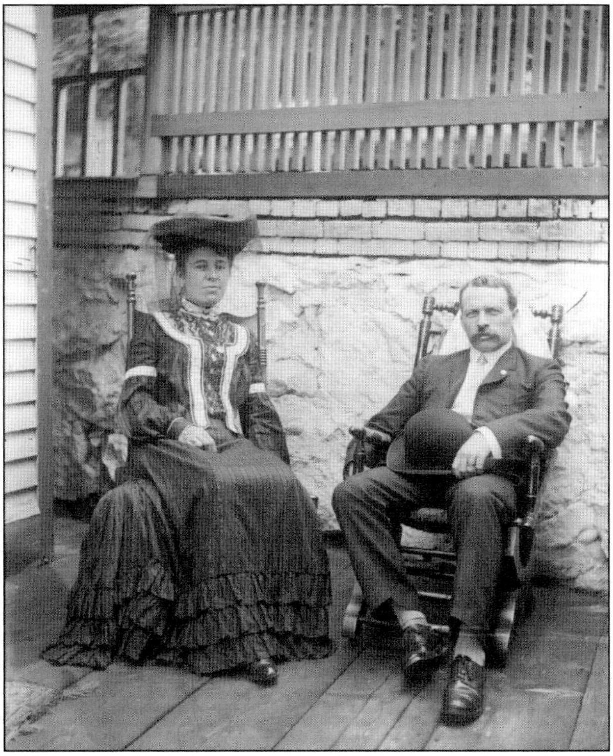

What is more pleasant on a sunny afternoon than sitting side by side on the back porch? In the late 19th century, formality was customary, indoors and out. In stark contrast to today, the lady wears a large hat, and the gentleman rests his bowler on his lap.

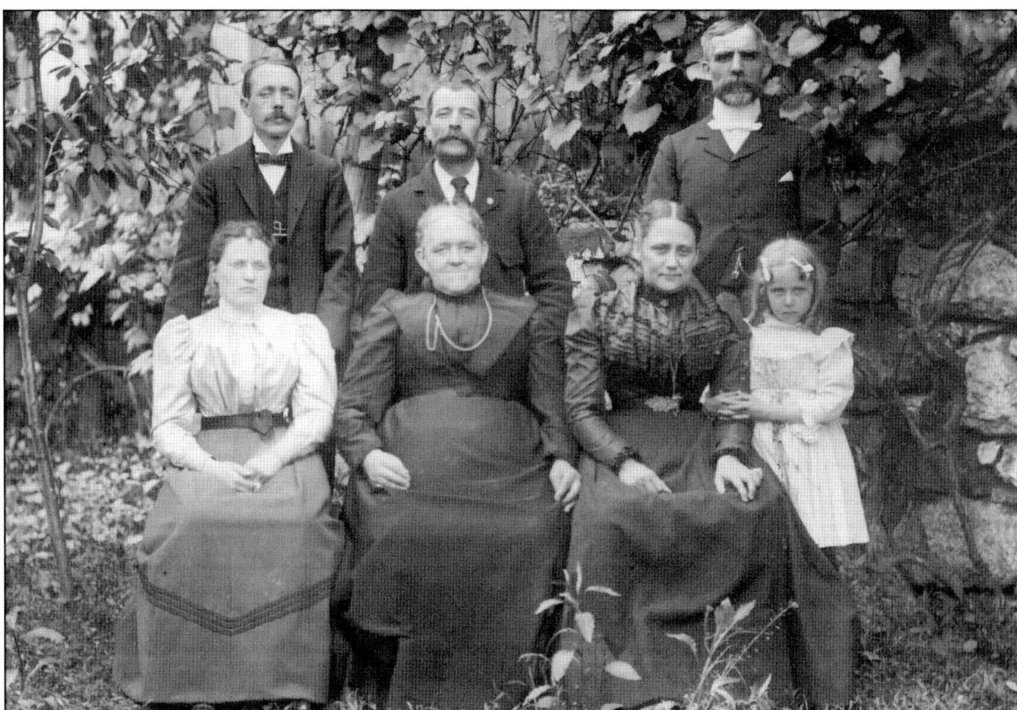

Does the absence from this photograph of the boy in the frilly shirt (page 90) suggest that he is the young person laid out in the front parlor?

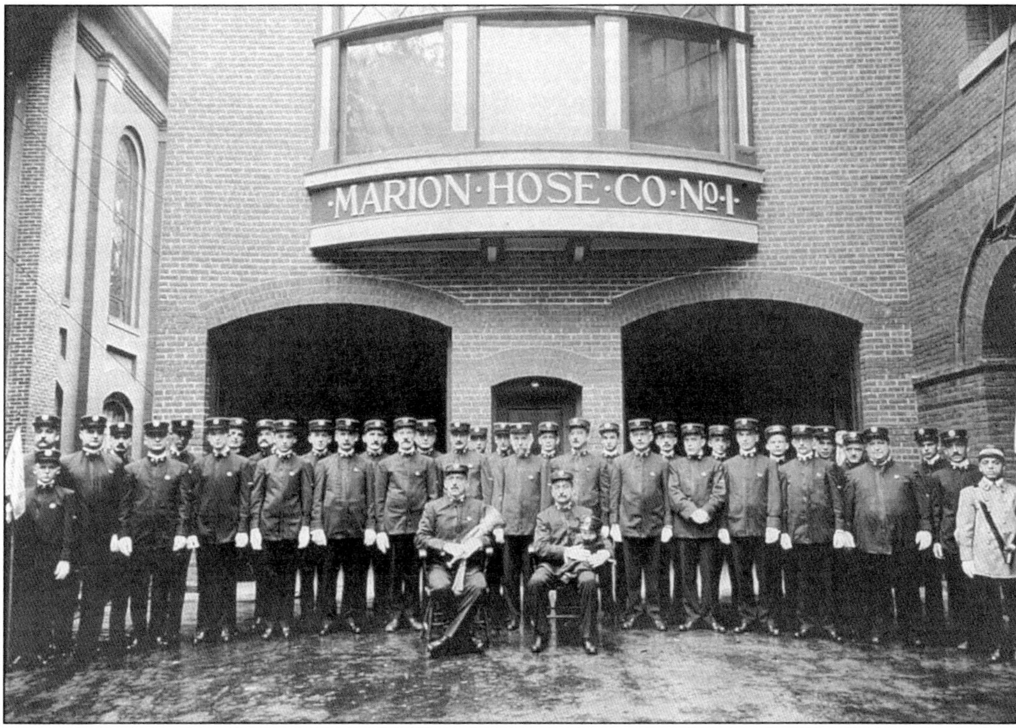

Following the devastating fire of 1849 that destroyed much of Race Street and Lower Broadway, the first firehouse in Mauch Chunk was built in 1853 to house the Marion Hose Company.

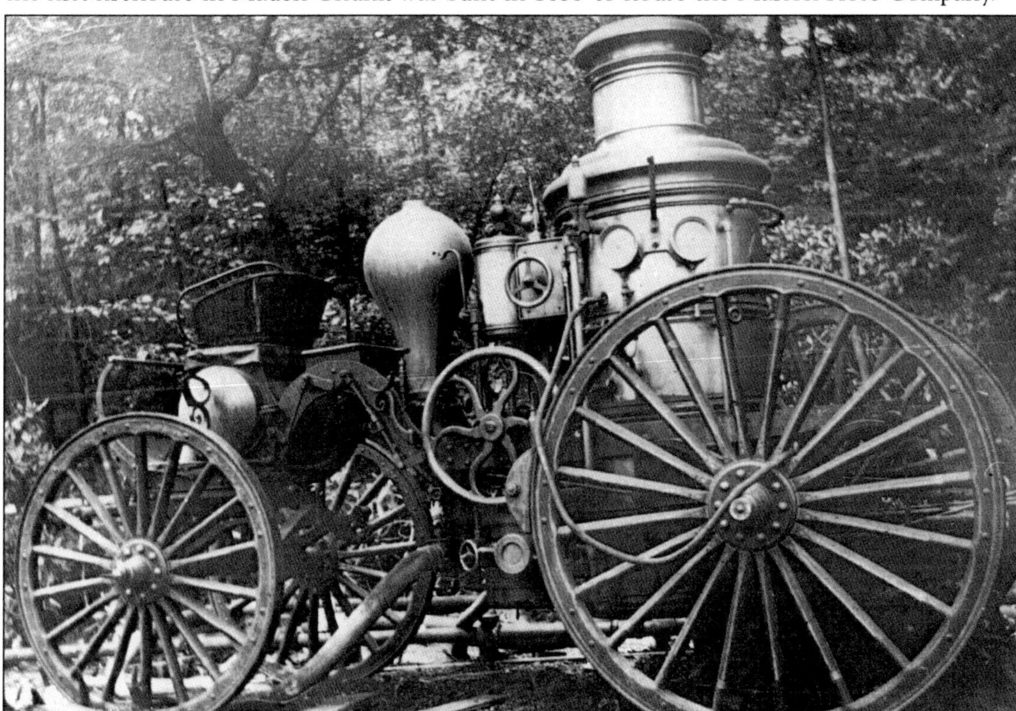
This was one of the earliest water pumpers used by the department.

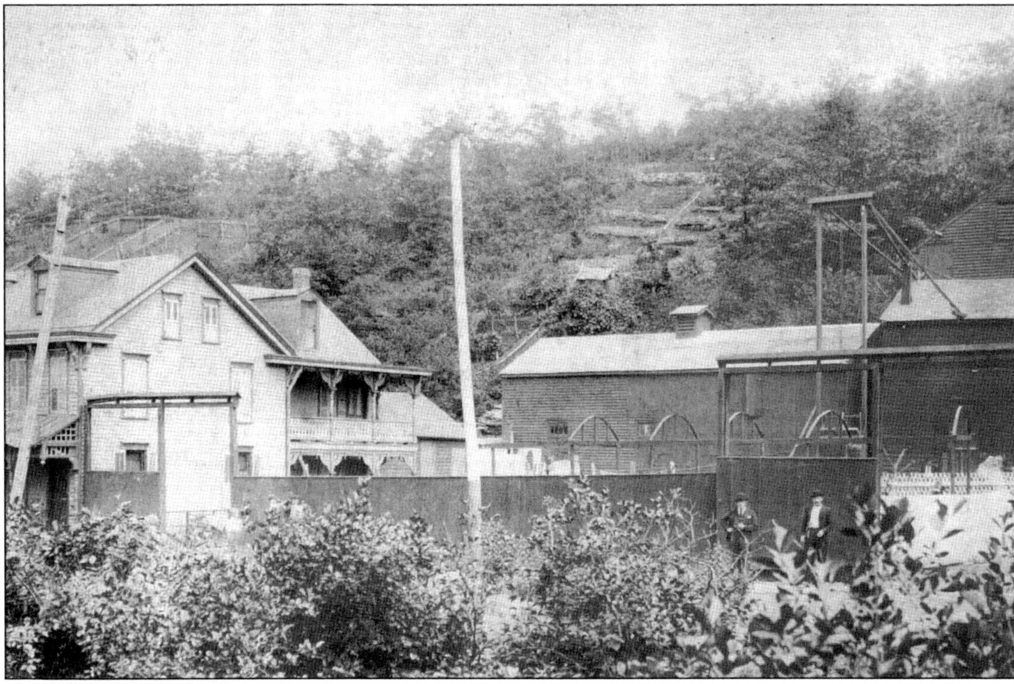

The West End Brewery was one of several breweries in town. (Courtesy Harold Queen.)

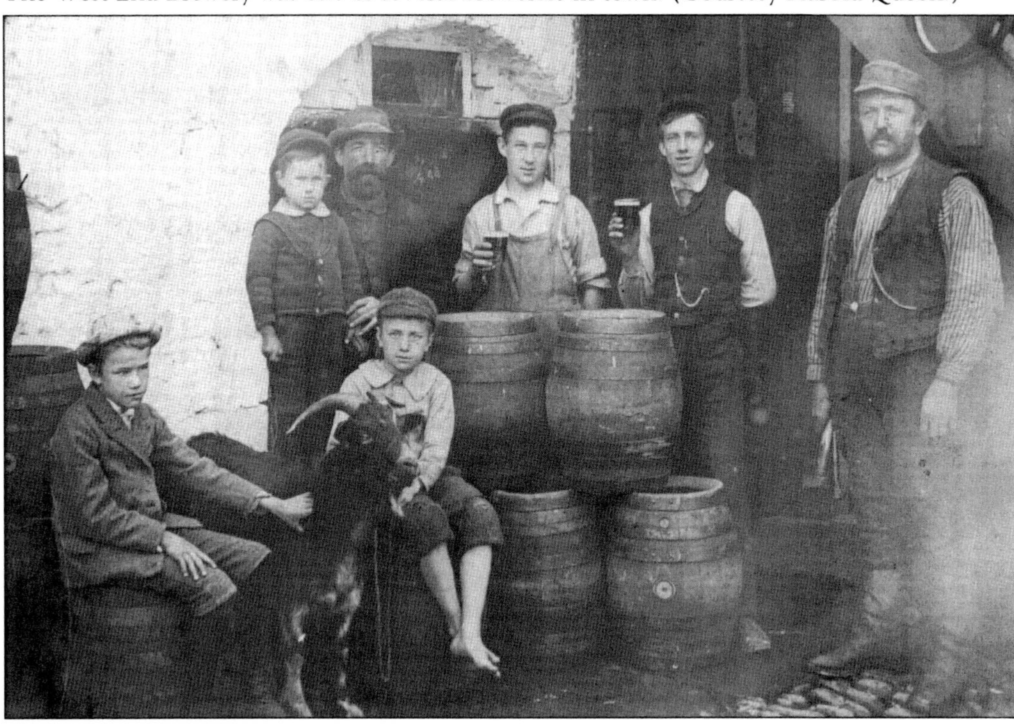

"When old Mauch Chunk was young, / Josiah used to say, / 'A man who labors hard / Should have six billy cups a day.' / And so, with an unsparing hand, / The whiskey flood was flung / But drunkards they were made by scores / When old Mauch Chunk was young." There was no shortage here. (Courtesy Harold Queen.)

The front stoop was a good place for a photograph. This diverse group poses outside the West End Brewery. (Courtesy Harold Queen.)

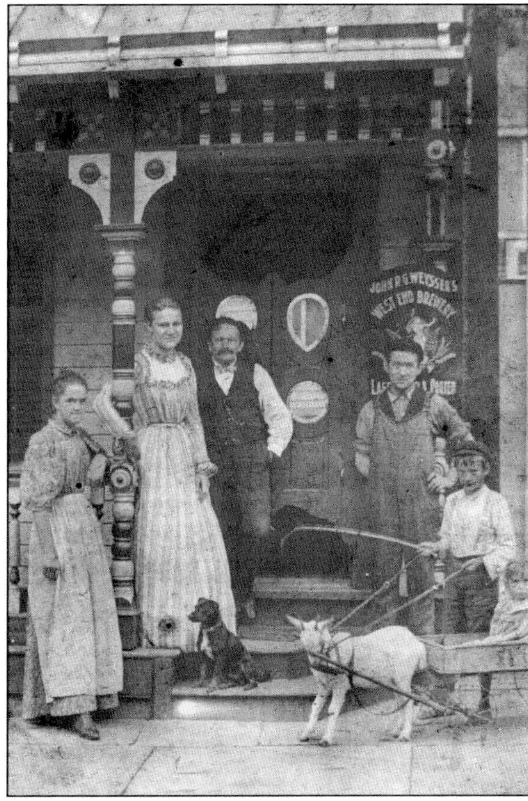

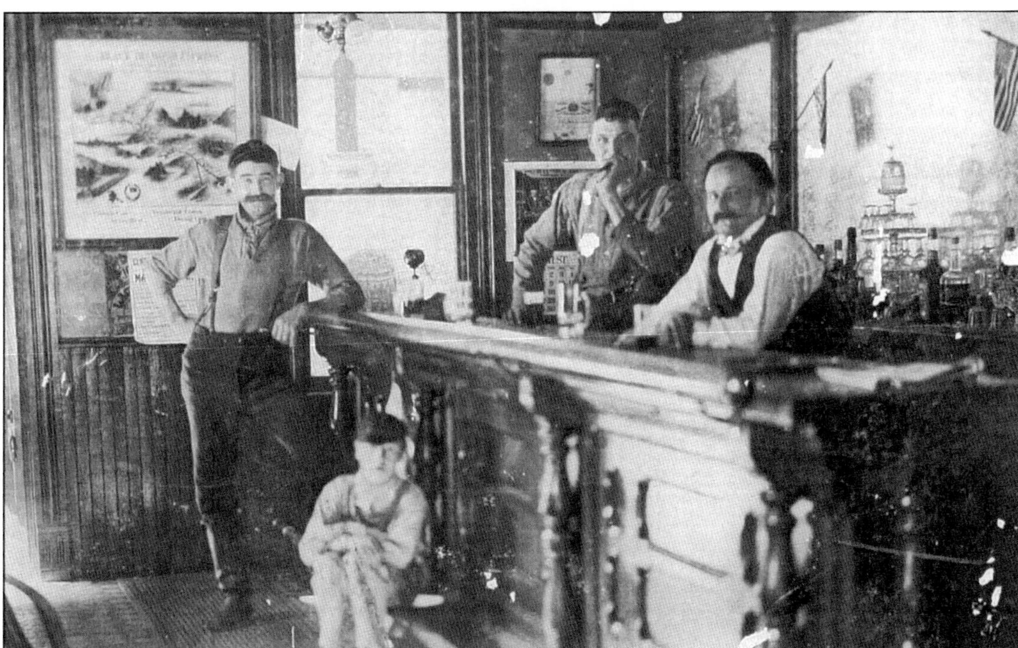

In the days before children were prohibited from entering drinking establishments, a young boy accompanies his father to the tavern. (Courtesy Harold Queen.)

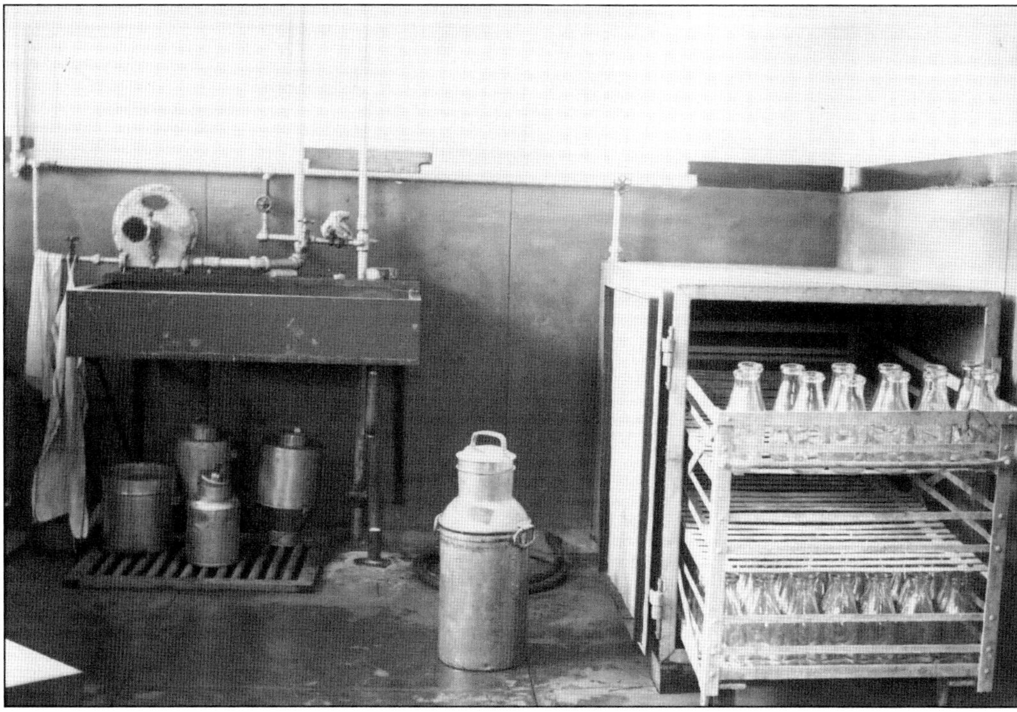

For teetotaling Mauch Chunkers, there was always milk from the local dairy.

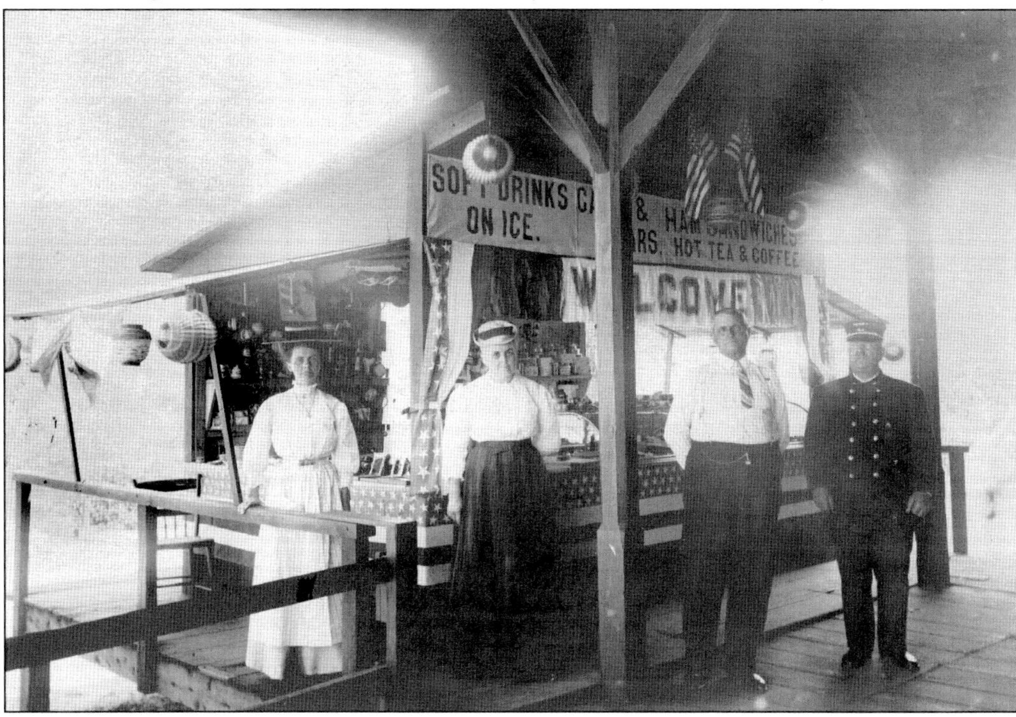

There was also Moxie at the Switchback beverage stand. Shown, from left to right, are Emily McAllister Johnson, Mary McAllister (Emily's mother), John McAllister (Emily's father), and an unidentified conductor. (Courtesy Beverly De Lay.)

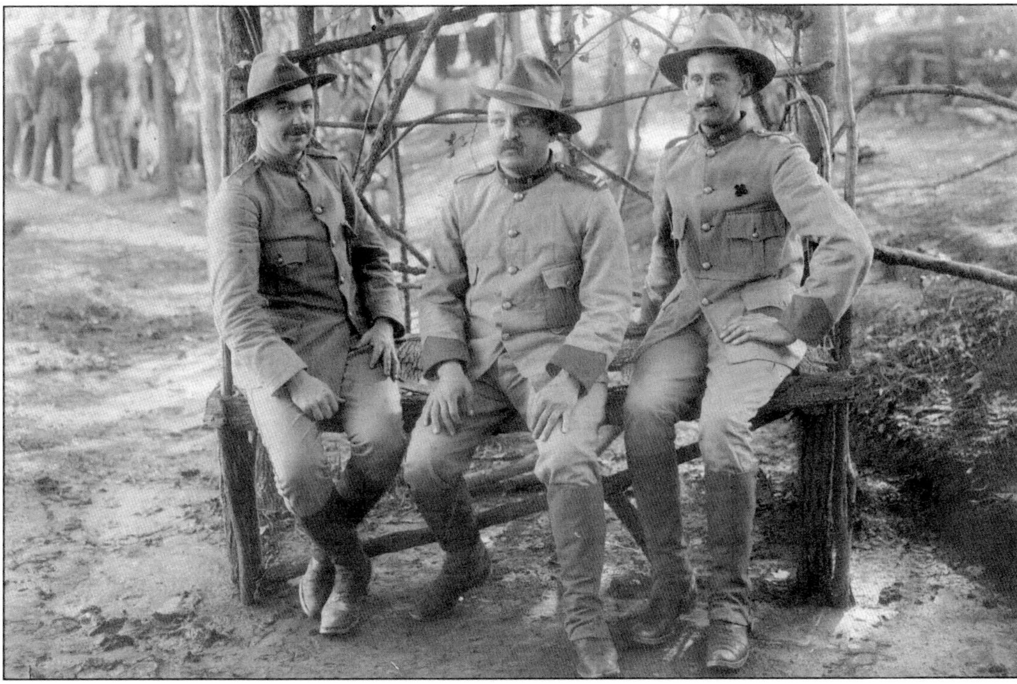

Posing in July 1898 are veterans of Company E, 1st Division, 3rd Brigade, 2nd Army Corps. They are, from left to right, Captain Harris, Lieutenant Britz, and Captain Blakslee.

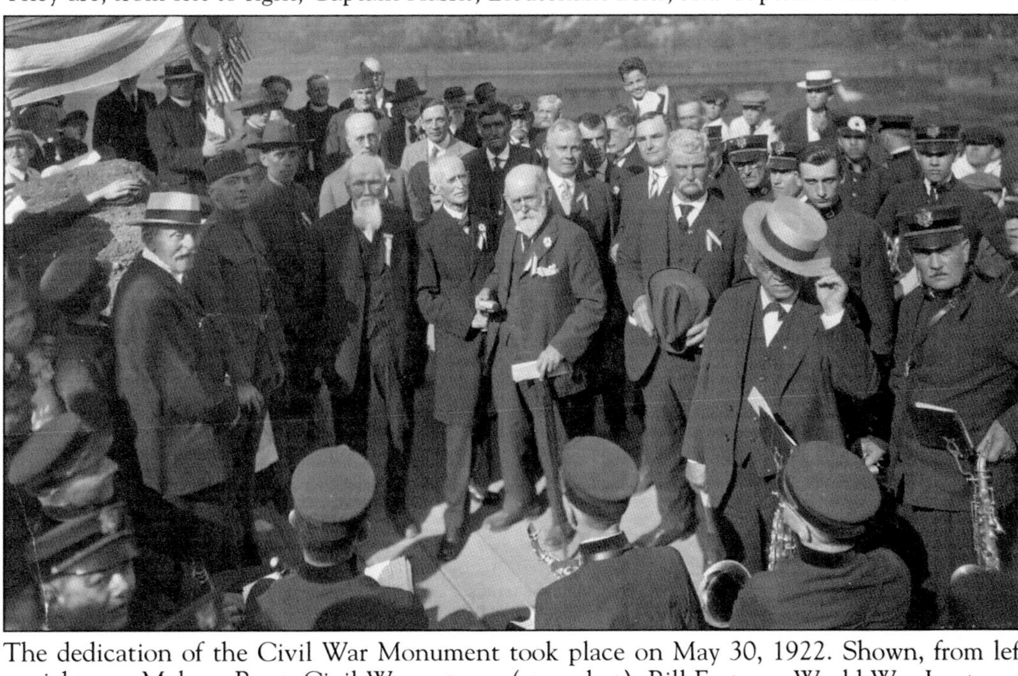

The dedication of the Civil War Monument took place on May 30, 1922. Shown, from left to right, are Malassa Root, Civil War veteran (straw hat); Bill Ecstone, World War I veteran; an unidentified veteran; Edward Mulhearn, attorney for the Mollie Maguires; Dr. Erwin, dentist; George McGovern; Mike Barret, Civil War veteran; Reverend Rehrig, Lutheran pastor (tipping hat); and Frank Kauffman, bandmaster of the Mauch Chunk Victory Band. (Courtesy Ray Holland.)

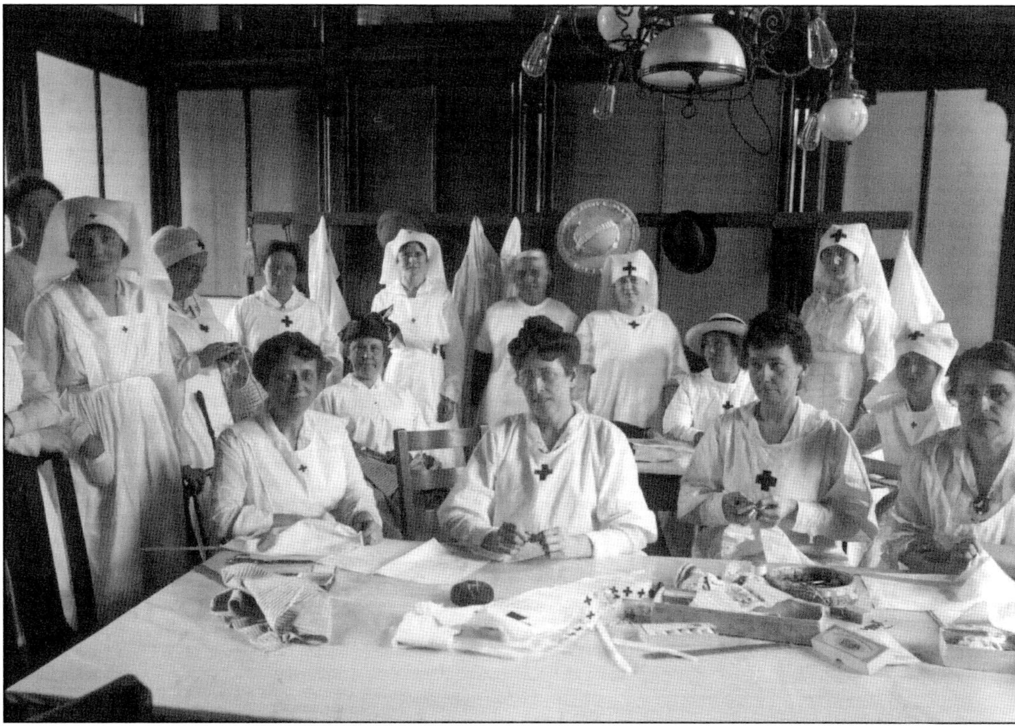

Red Cross workers volunteered their time during World War I.

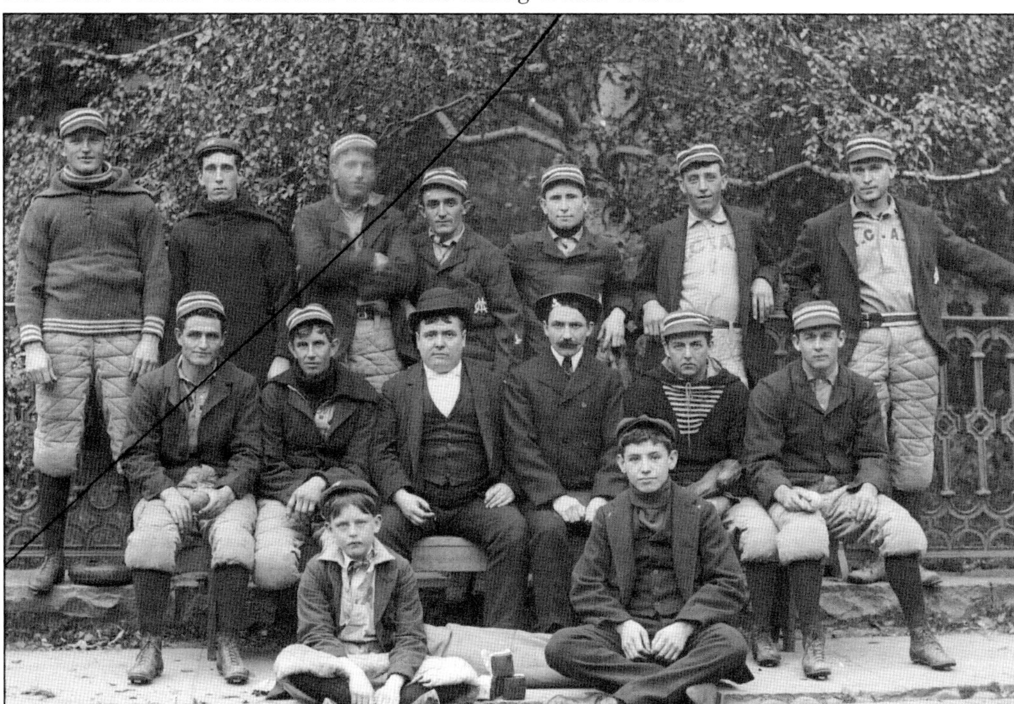

Sport was an important part of Mauch Chunk life. This *c.* 1900 photograph shows the Mauch Chunk Athletic Association.

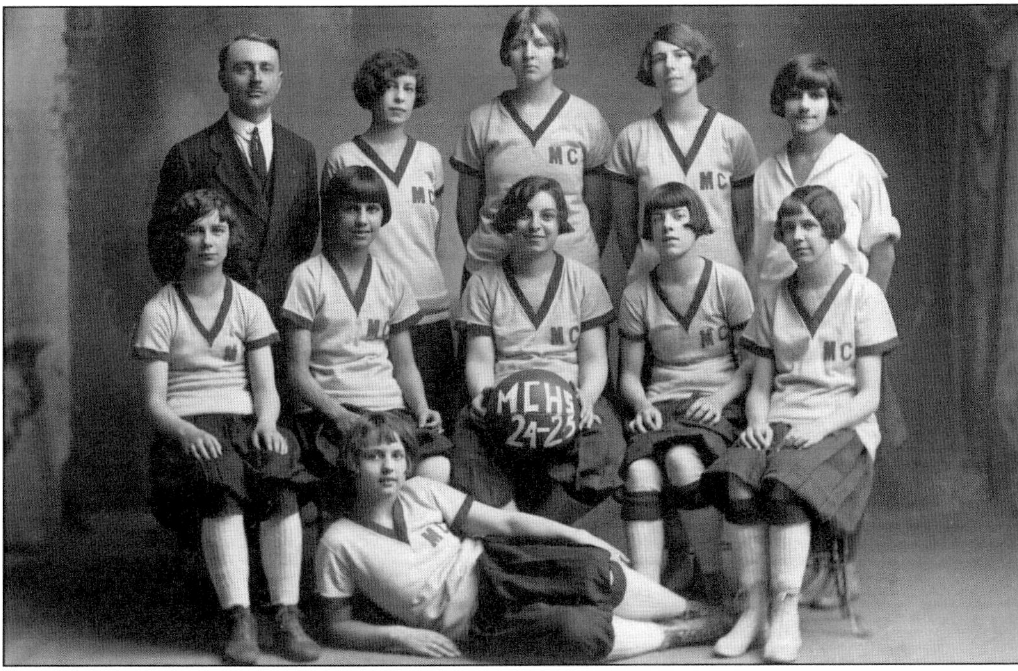

Mauch Chunk High School had its share of champions. Here, the 1924–1925 girls' basketball team poses for an end-of-season photograph. (Courtesy Virginia Wells.)

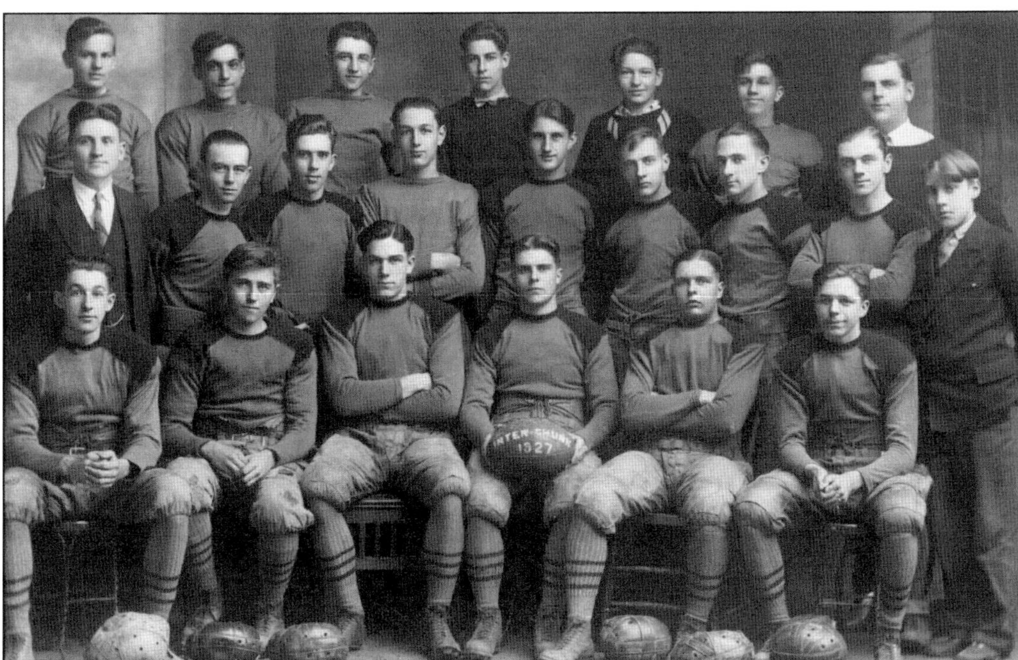

The high school was proud of its 1927 football champions. (Courtesy Virginia Wells.)

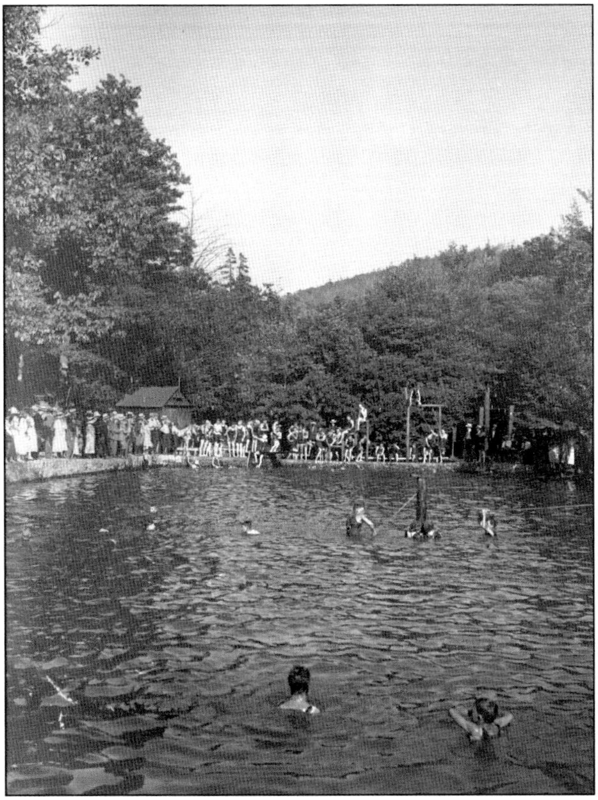

A swimming pool was built on the Lentz Trail above Mauch Chunk at Flagstaff Junction. Equipped with tables and benches, it was a shady spot for tourists and Mauch Chunkers to cool off on a hot summer day. Trolleys ran between Flagstaff, points of interest in Mauch Chunk, and the railroad stations.

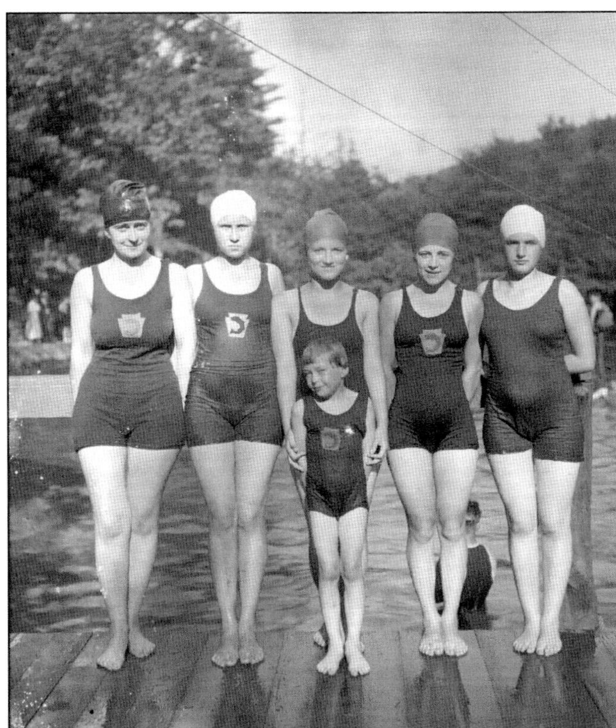

These Mauch Chunk bathing beauties pose after cooling off on a hot summer day.

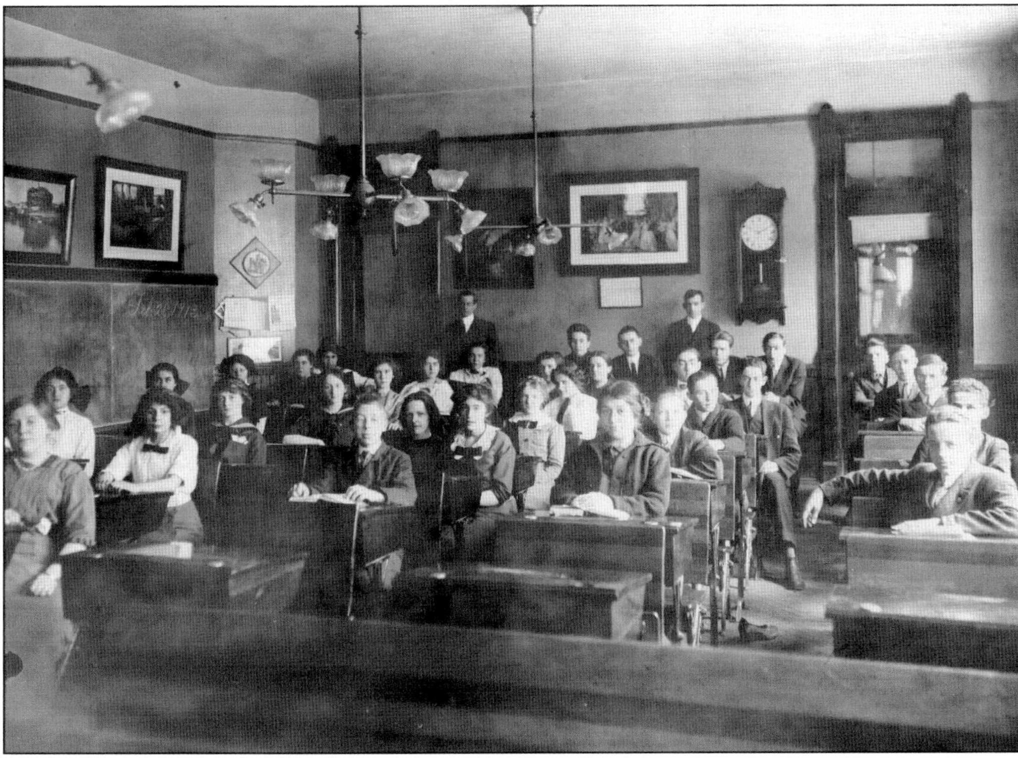

The classroom was sparse and the desks hard, but Mauch Chunk had its share of scholars. Miss Deatrich (left front), Mr. Emery, and Mr. Werner pose with a 1913 class at Mauch Chunk High School.

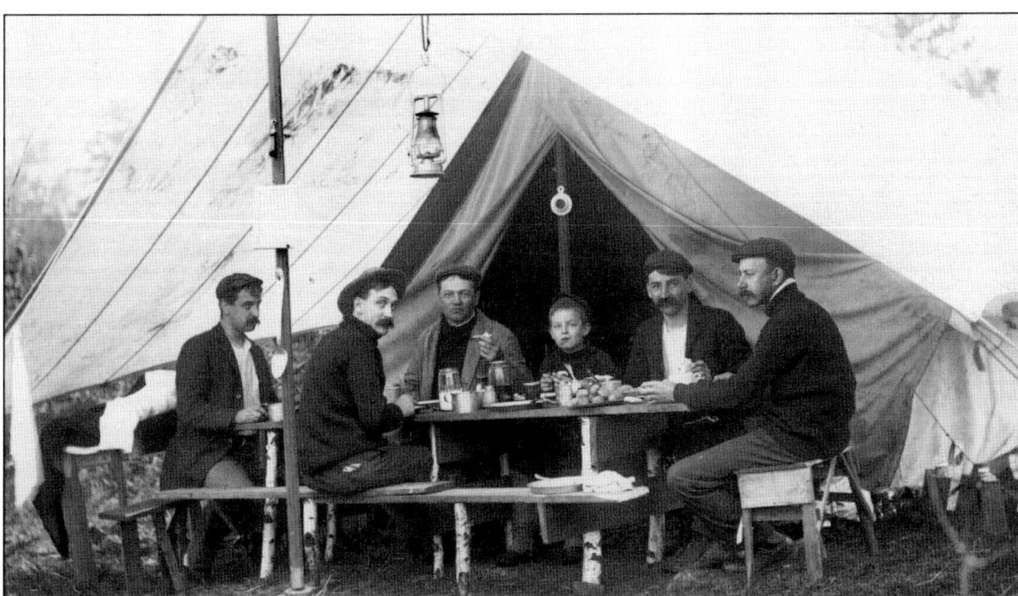

Campers enjoy a meal at Big Pond (now Lake Harmony) in September 1901.

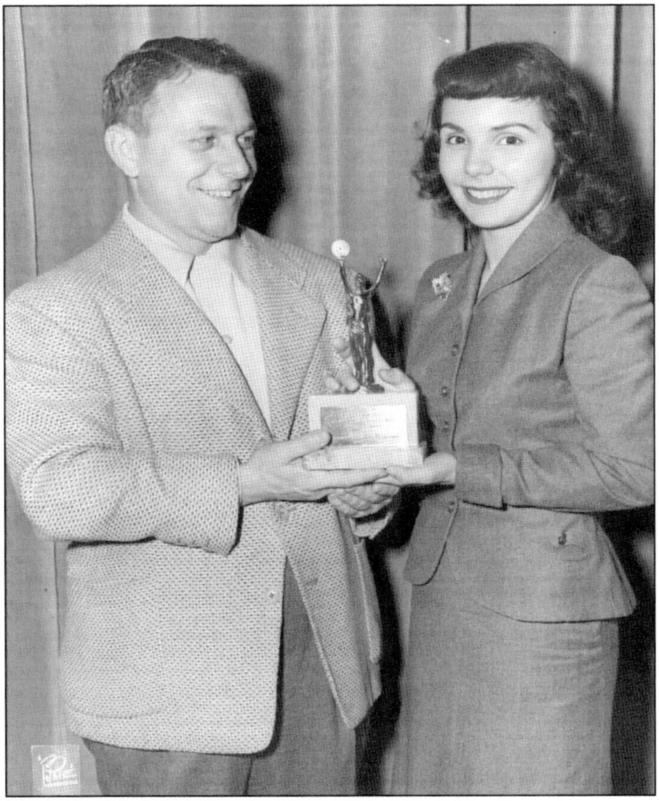

Singing sensation Teresa Brewer is shown with her manager, Summit Hill native Richie Lisella. Brewer's first professional engagement was at the American Hotel (now the Inn at Jim Thorpe). Richie Lisella was road manager for the Dorsey Brothers Band and Frank Sinatra. (Courtesy Richie Lisella.)

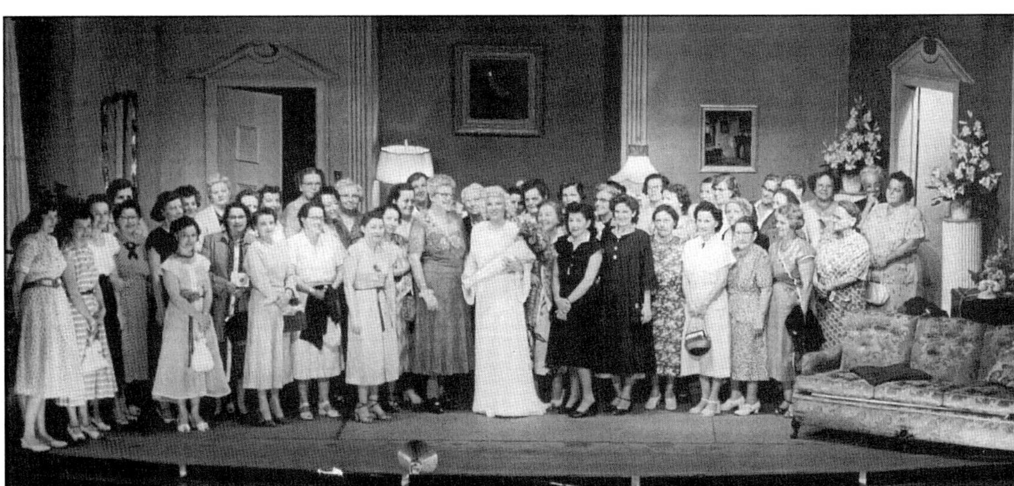

Mae West (center) was one of the notables who performed at the Mauch Chunk Opera House. (Courtesy Tom Knappenberger.)

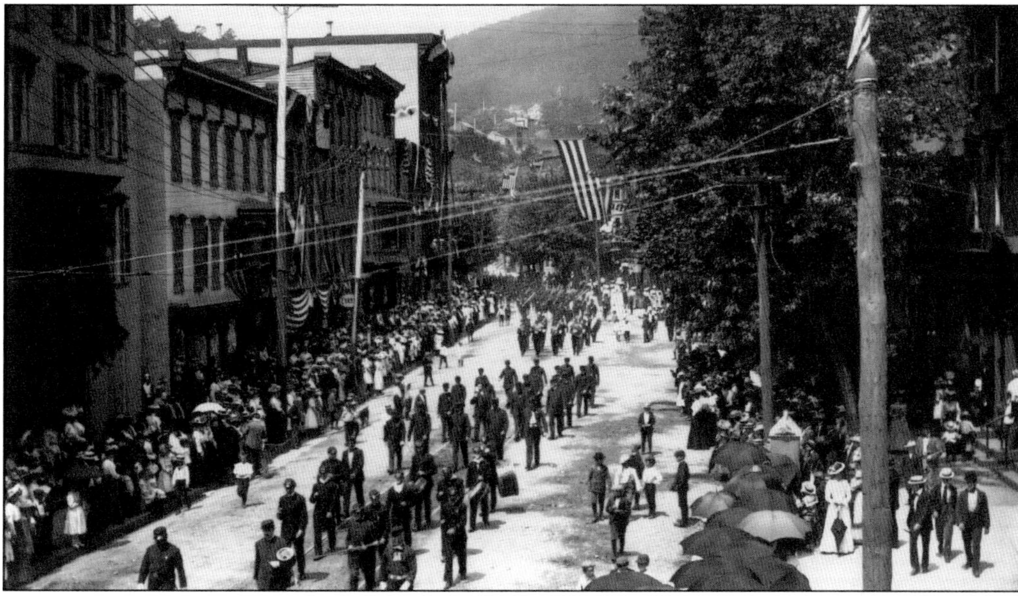

Following the unfurling of the giant flag across the Lehigh River on July 4, 1898, the parade makes its way down Broadway to Packer Park. (Courtesy Tom McBride.)

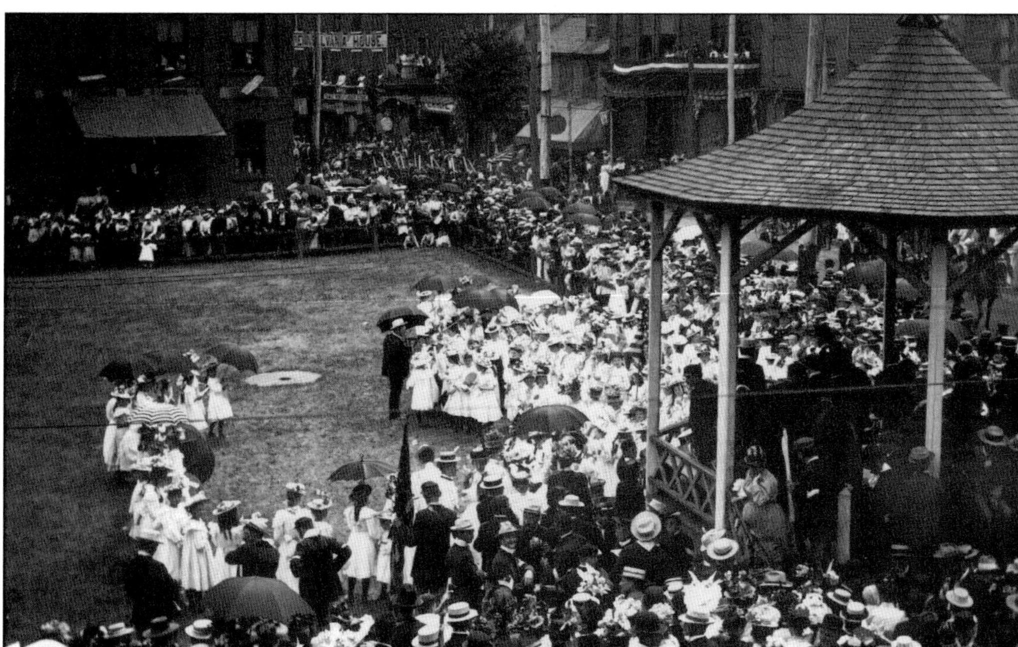

Children entertain the crowd in Packer Park. (Courtesy Tom McBride.)

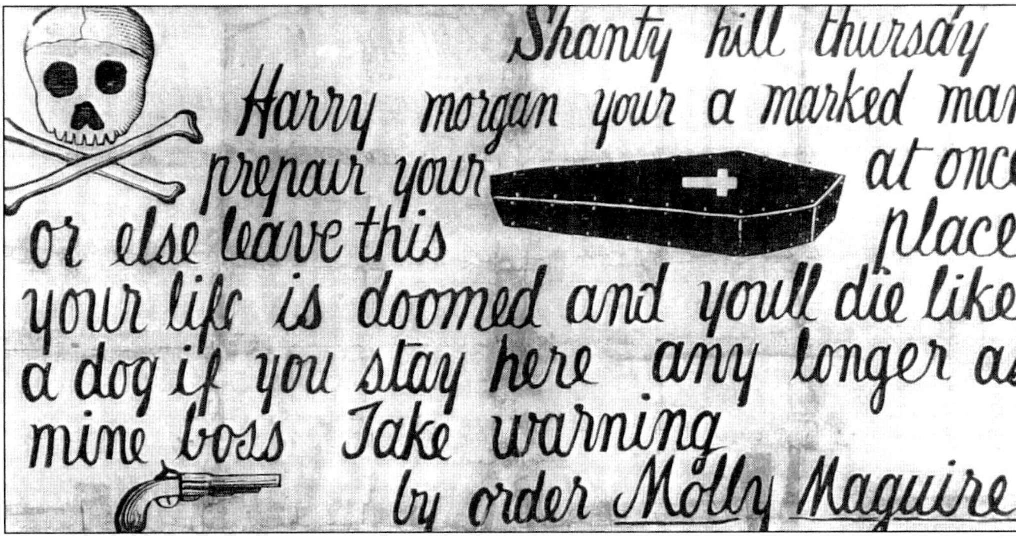

While tourists marveled at the scenery and Mauch Chunk prospered, unrest in the coal fields smoldered like a mine fire. As the coal bosses grew richer and immigrant workers slaved in the mines, resentful Irishmen formed the Molly Maguires to wage war against victimization and injustice. Warnings, emblazoned with crude drawings of coffins, skull and crossbones, and pistols were directed at targeted individuals. Failure to comply with the Mollies' demands meant almost certain death.

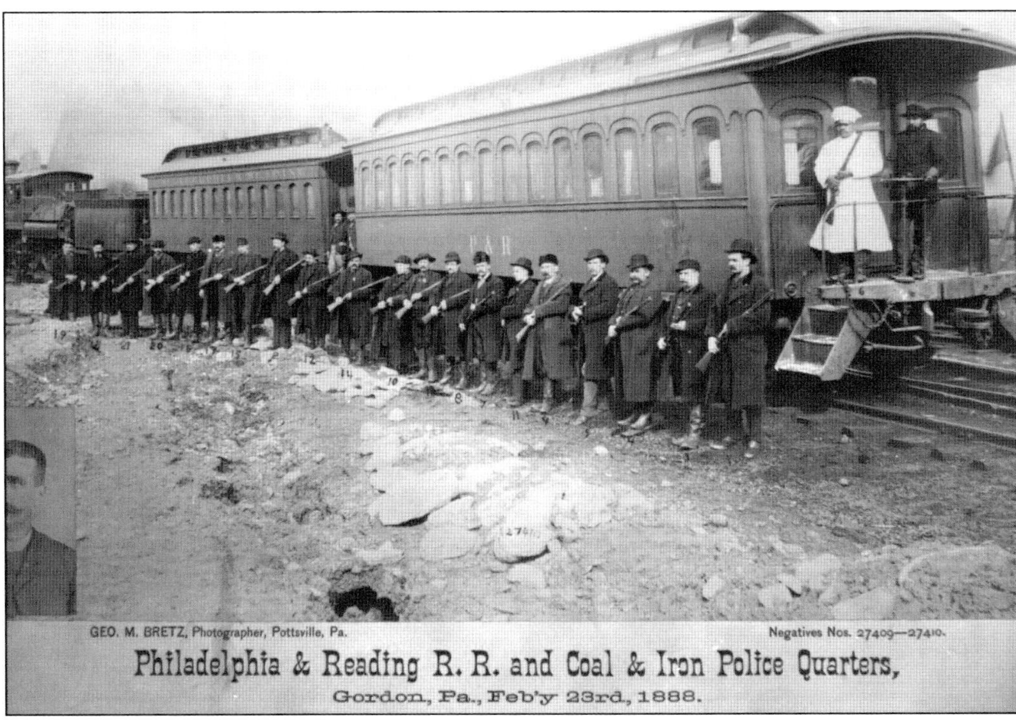

In 1873, 28-year-old Pinkerton detective James McParlan (alias James McKenna) was hired by the Reading Railroad to penetrate the Shenandoah Division of the Hibernians by becoming a member of the organization. As a result of McParlan's undercover work, 20 Mollies were hanged in Carbon and neighboring counties.

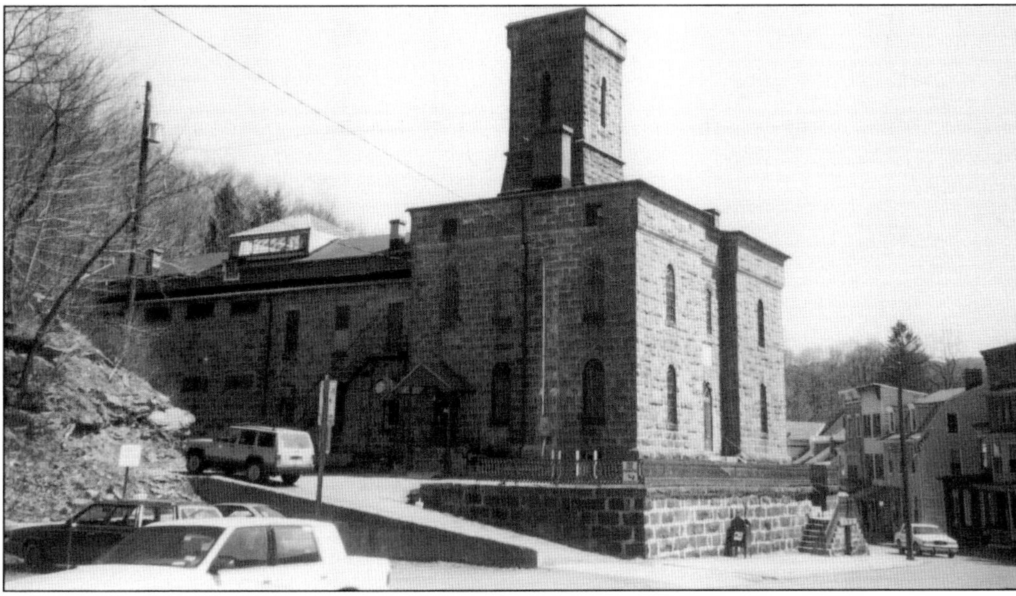

The Carbon County Jail is now the Old Jail Museum on West Broadway. Hanged in the jail on "Black Thursday," June 21, 1877, were Michael J. Doyle and Edward Kelly for the 1875 murder of mine foreman John P. Jones; John Donahue for the 1871 murder of Morgan Powell, assistant superintendent of the Lehigh & Wilkes-Barre Coal & Iron Company; and Alexander Campbell for the murders of both Jones and Powell. On March 28, 1878, Thomas P. Fisher was hanged for the murder of Morgan Powell and, on January 14, 1879, James McDonnell and Charles Sharp were hanged for the murder of George K. Smith, coal baron and owner of the George K. Smith & Company.

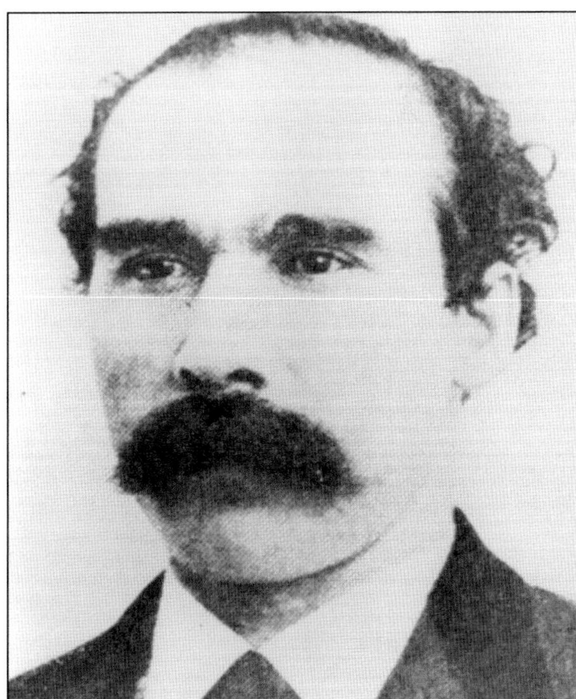

As Alexander Campbell, shown here, was taken out of cell 17 to be hanged, he placed his left hand on the wall, leaving a hand print. As he viewed his mark, he swore that it would remain on the wall forever as testimony to his innocence. To this day, neither cleaning nor painting has been able to remove Campbell's hand print.

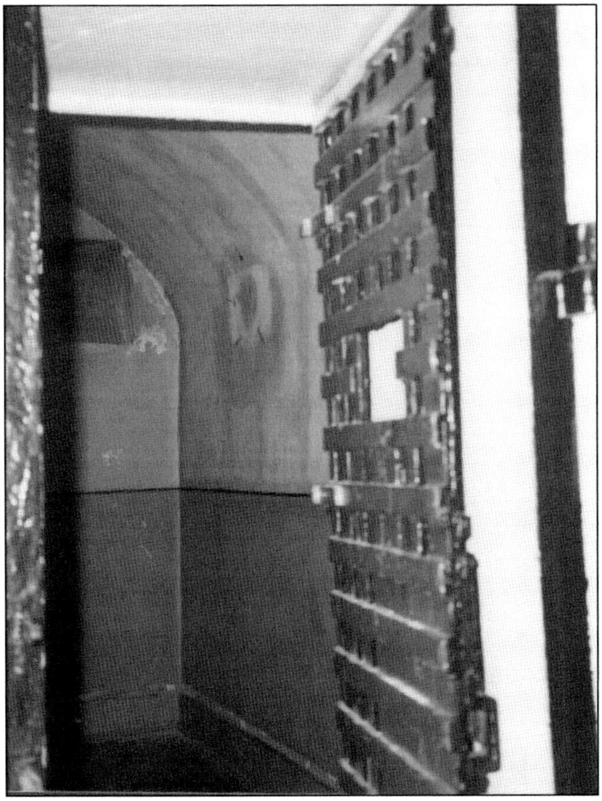

Alexander Campbell's hand print remains on the wall in cell 17. Following the hangings, the Irish Catholic Society labeled the Mollies thugs and expelled members of the Ancient Order of Hibernians from the church. In the community, help wanted signs carried the words, "No Irish need apply."

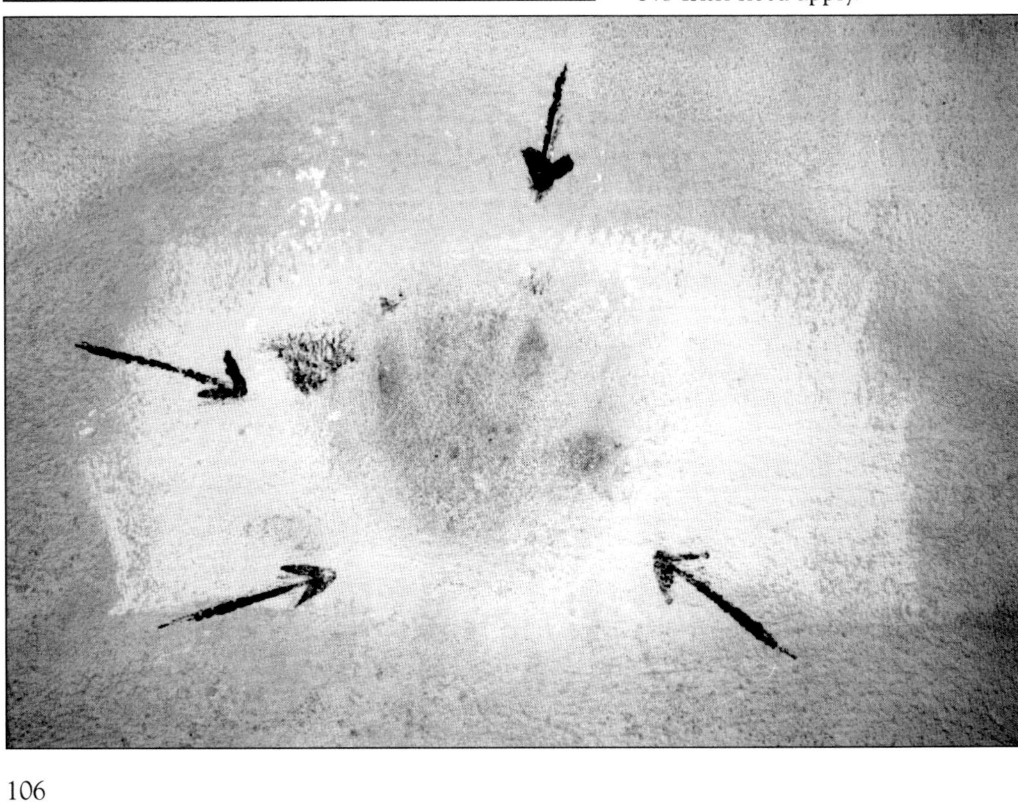

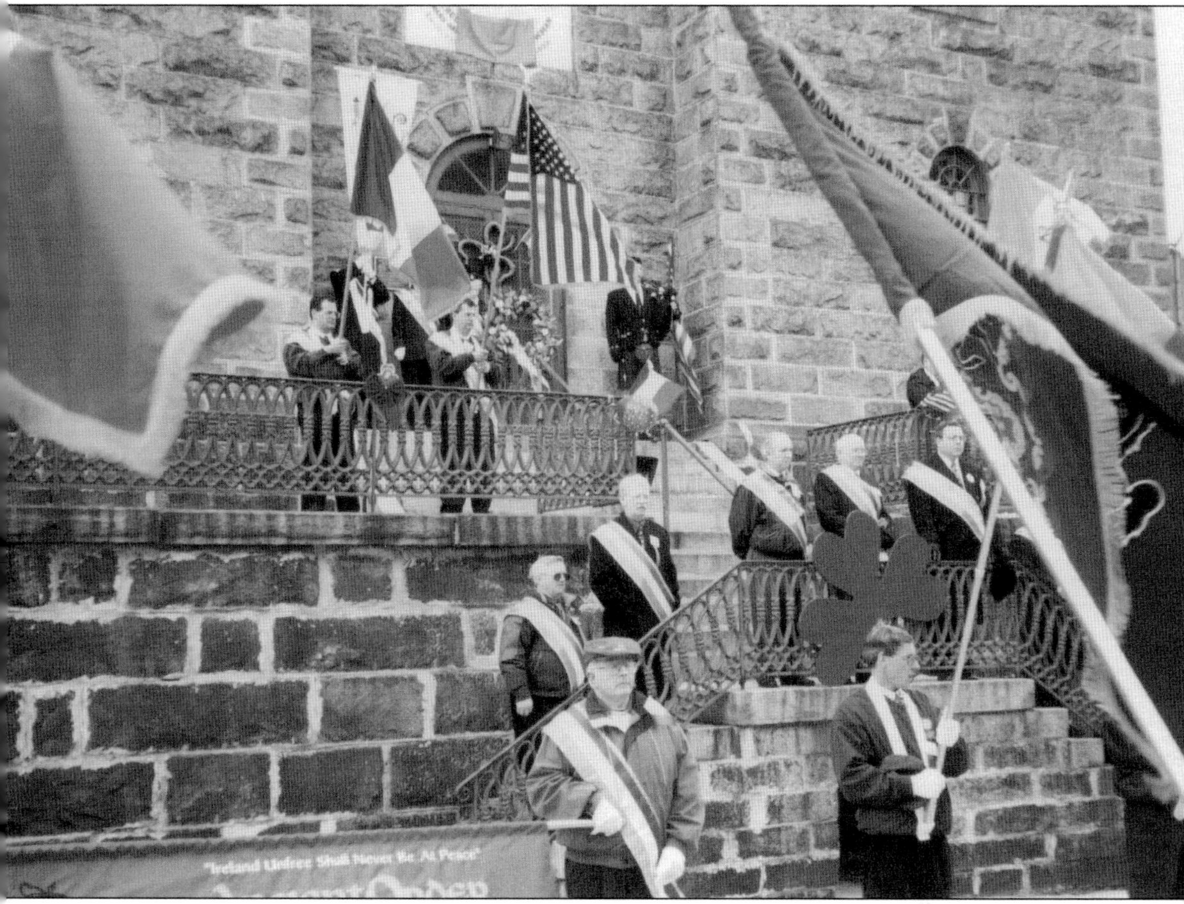

Today in Jim Thorpe, the Mollies are seen through the eyes of revisionist history. They are no longer labeled thugs but are regarded as men who fought for fair labor practices. As chapters of Hibernians from all over the area proudly bear their standards down Broadway each St. Patrick's Day, they pause outside the jail and remember their hanged brethren with a prayer. (Courtesy Anne Baddick.)

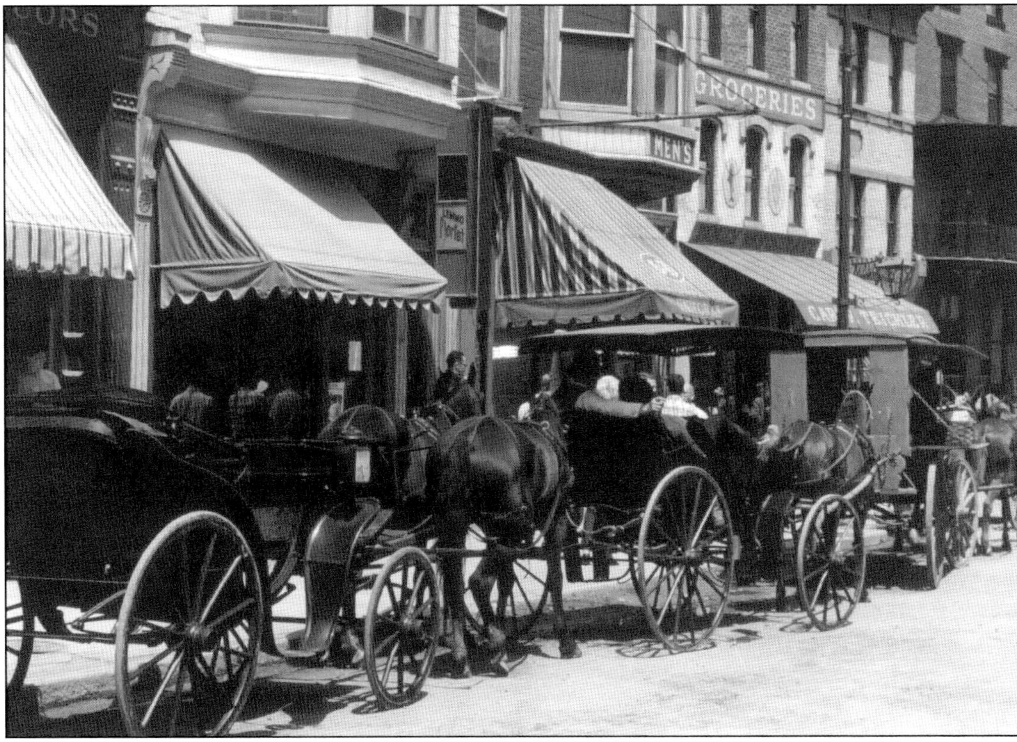

The town of Jim Thorpe goes Hollywood. Paramount Pictures' *The Molly Maguires* was filmed in Jim Thorpe and the coal fields in 1968. During the shoot, buildings and storefronts took on a look of the late 19th century, and Broadway became a dirt road. (Courtesy Worthington Heimbach, Bethlehem Photograph Color Club.)

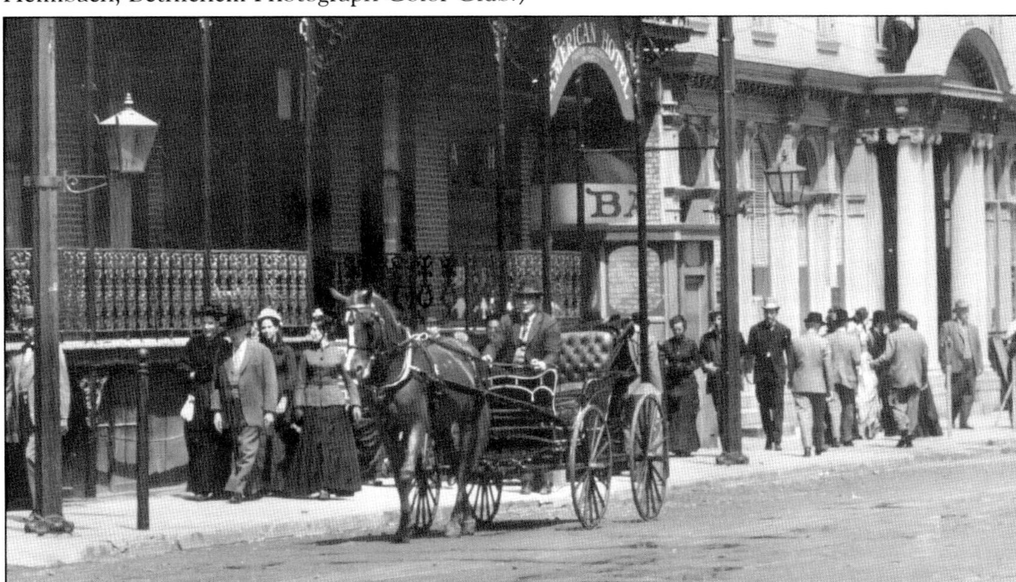

No set had to be constructed for the filming of *The Molly Maguires*. Jim Thorpe's Victorian buildings stood tall and proud, ready for the action to begin. A horse and carriage are not out of place in front of the American Hotel, now the Inn at Jim Thorpe.

Three

Decline

By 1920, petroleum had dealt Mauch Chunk a double blow. It was taking the place of coal in home and industry while fueling the popular new automobile. As petroleum replaced coal, canal payloads were pirated by railroads, and rail excursion tourism was taken over by trips in the automobile, one by one, all the things that made Mauch Chunk prosperous closed down.

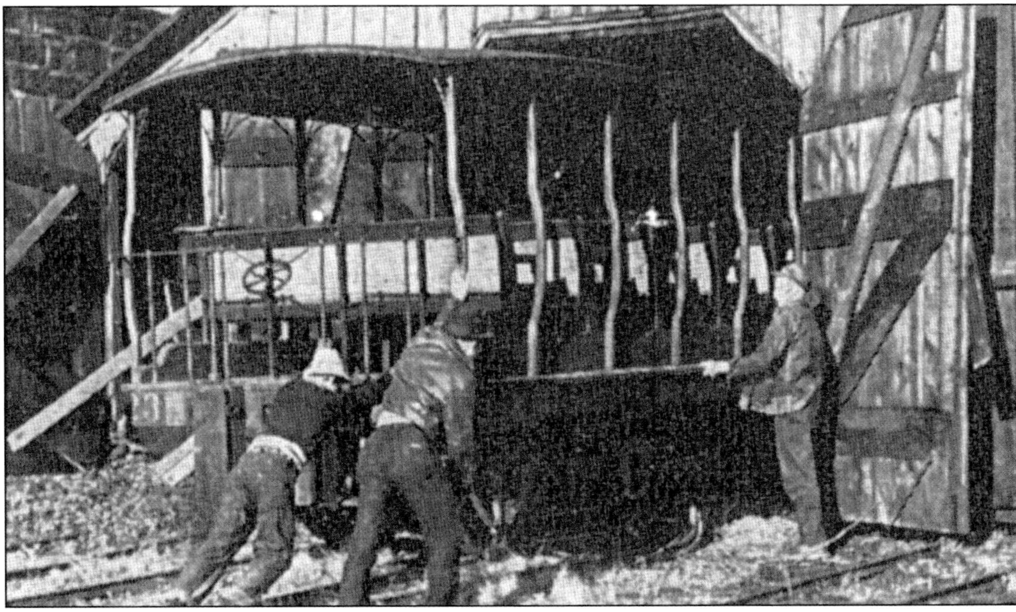

The Switchback ceased operating in 1933 when the main drive shaft bearing in the Mount Pisgah Engine House broke. On September 27, 1937, 18 miles of track, 11 passenger coaches, 3 flatcars, a tank car, powerhouses, carbarns, and repair shops were sold at auction. The scrap iron was sold to Japan. When war broke out in the Pacific, Mauch Chunkers remarked bitterly that their famous Switchback was coming back as bullets.

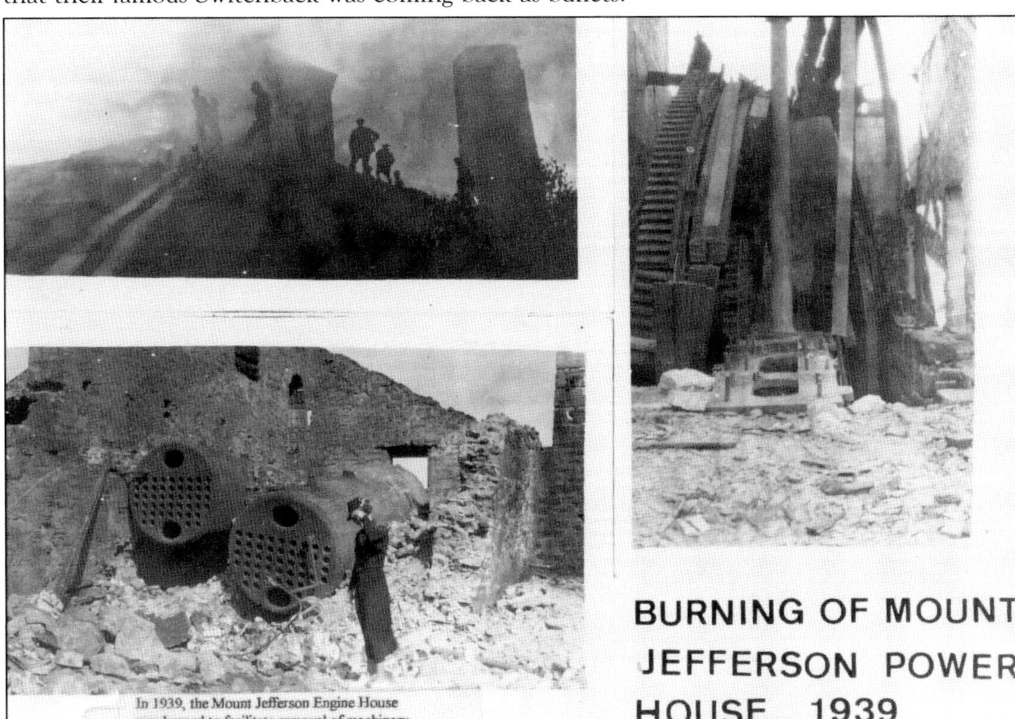

The Mount Jefferson Power House was burned to facilitate removal of machinery. (Courtesy Agnes McCartney.)

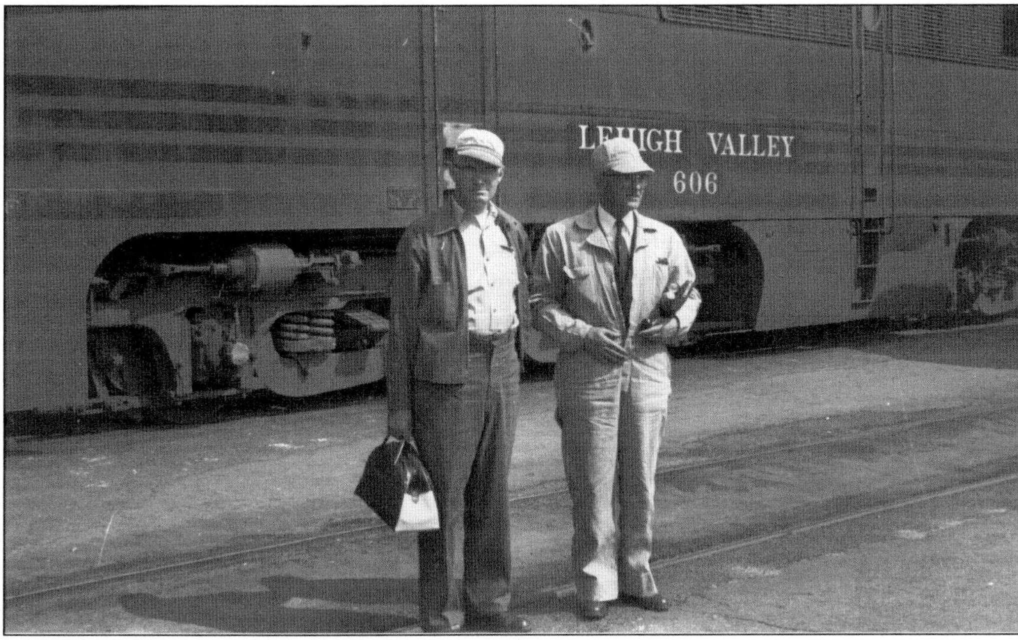

The engine crew of the last westbound No. 9, the "Black Diamond," is shown at Lehighton, Pennsylvania, on May 11, 1959. On a snowy February 4, 1961, passenger traffic through Mauch Chunk ceased when the No. 8, "Maple Leaf," made its last run from Buffalo to New York. The Lehigh Valley Railroad's past glory made its end all the more humiliating. On March 30, 1976, 125 years after it began operating, it went broke. "The most beautiful train in the world" was history. (Courtesy Ray Holland.)

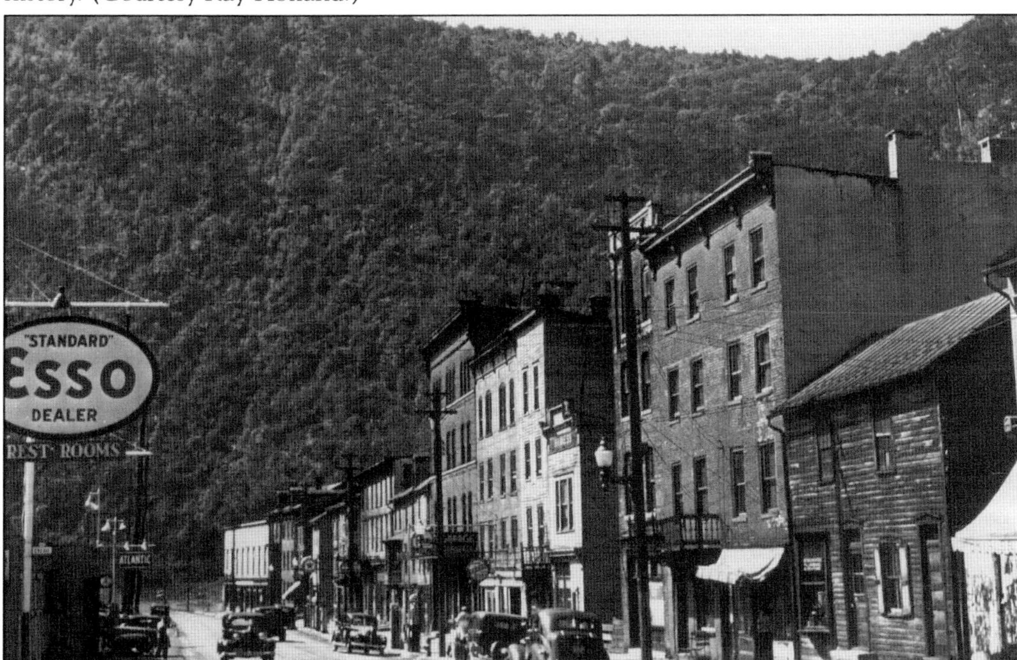

Visible in this Depression-era photograph, when oil and gasoline figured so prominently in Mauch Chunk's demise, are the Esso and Atlantic gas station signs and vintage automobiles.

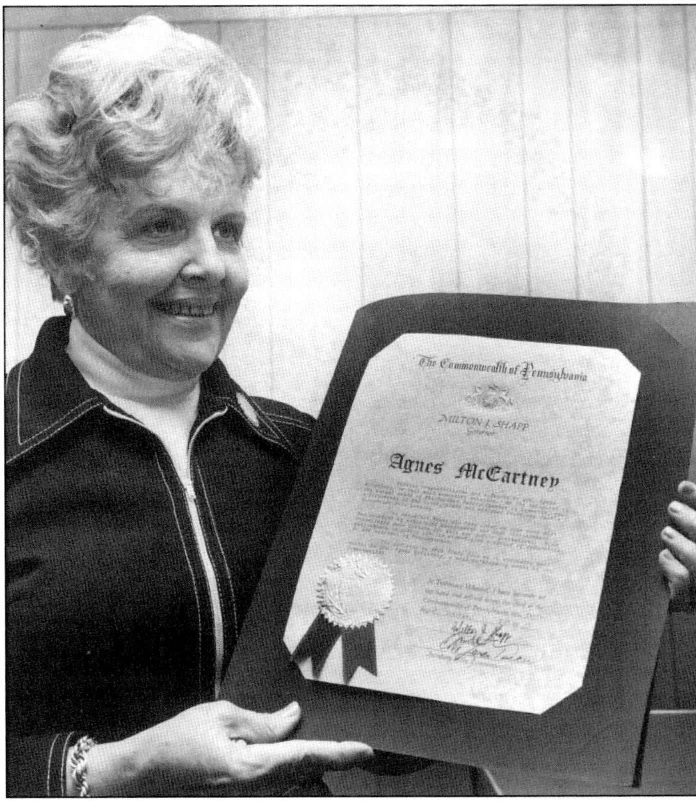

As Mauch Chunk struggled to survive, foremost among the visionaries were Agnes McCartney and Joe Boyle. McCartney's action came through the Carbon County Tourist Promotion Agency, which she headed for 30 years. Boyle worked to promote Mauch Chunk through the newspaper he inherited from his father, the *Mauch Chunk Times-News*. Shown here is Agnes McCartney, director of the Carbon County Tourist Promotion Agency and the county planner.

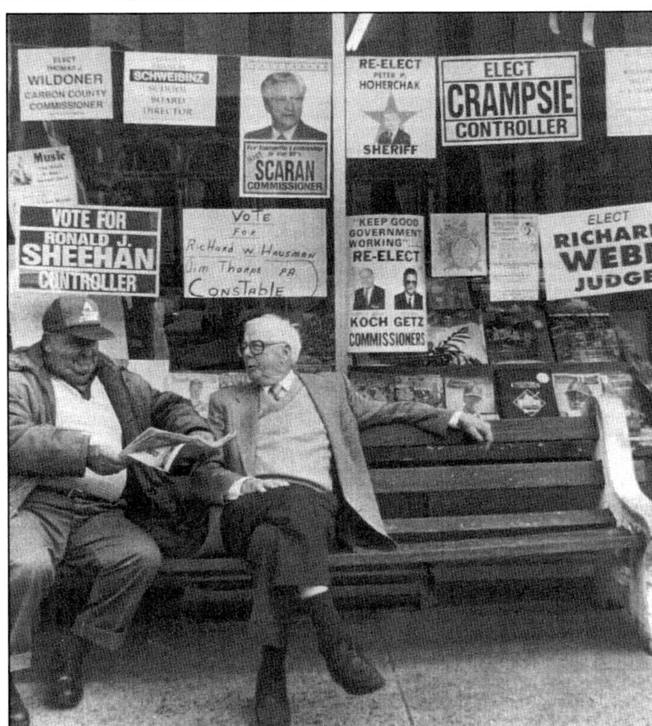

Joe Boyle (1915–1992) is shown on his usual bench in the center of town, conversing with locals and visitors.

Joe Boyle started a nickel-a-week campaign through his newspaper: the contributions of local families would be used to attract industry and revive tourism. Media headlines about Boyle's effort attracted the attention of Jim Thorpe's widow, Patricia. She was looking for a town that would honor her late husband by taking his name and constructing a mausoleum befitting his fame as an Olympian. Stripped of his gold medals for playing minor league baseball, Jim Thorpe had been buried in an unmarked grave in Tulsa, Oklahoma. In a *quid pro quo* scheme designed to attract business, rekindle tourism, and provide Jim Thorpe with a suitable resting place, it was proposed that Mauch Chunk and East Mauch Chunk unite to form the town of Jim Thorpe. (Courtesy Tom Potter.)

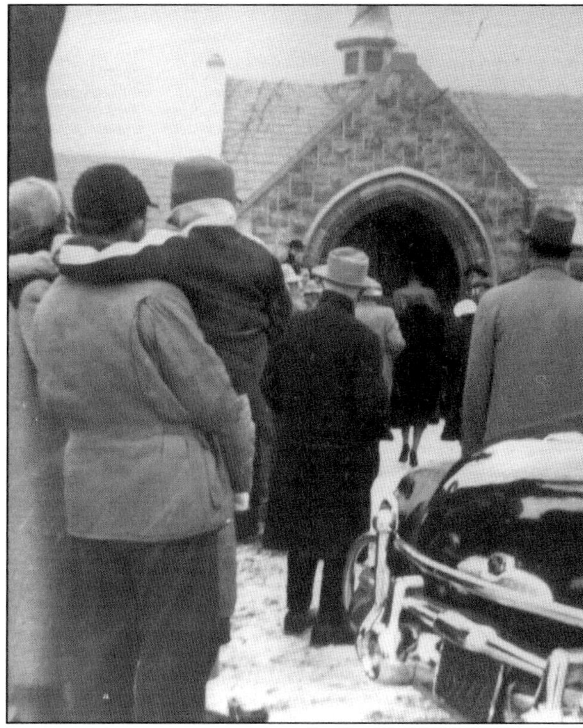

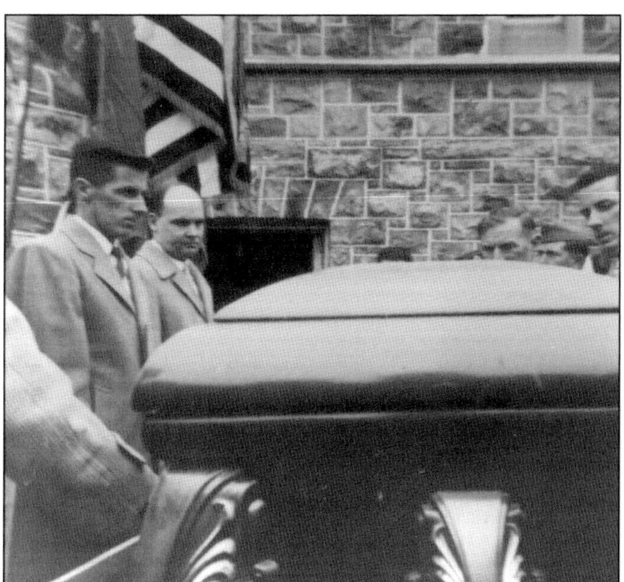

On February 8, 1954, the body of Jim Thorpe arrived from Oklahoma and was placed in a temporary crypt in Evergreen Cemetery until a mausoleum was constructed. When a pallbearer commented that the casket was so heavy that it must be filled with rocks, Mauch Chunkers took him literally. To squelch rumors that the town had been duped, Joe Boyle and two morticians opened the coffin. "We opened up the casket," Boyle recalled. "There was a plastic bag over the head. We removed it. There was no doubt it was Jim."

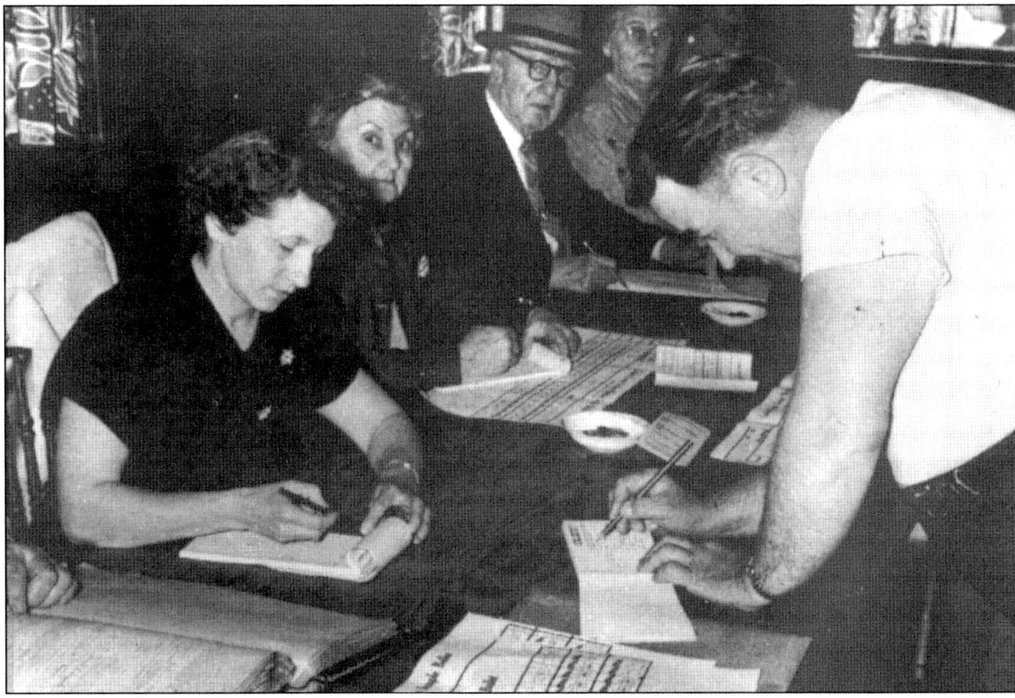

On May 18, 1954, a referendum to change the name of the town was placed before the voters. Here, a voter signs up to cast a ballot.

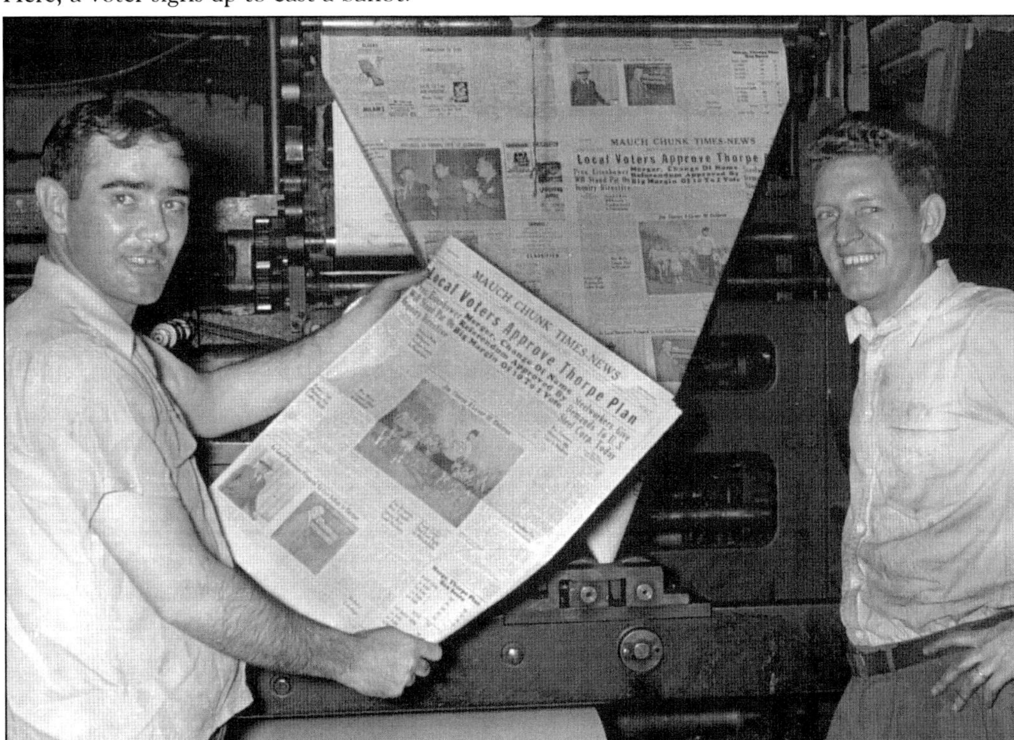

The *Mauch Chunk Times-News* proclaims victory. The referendum passed by a margin of 10-1.

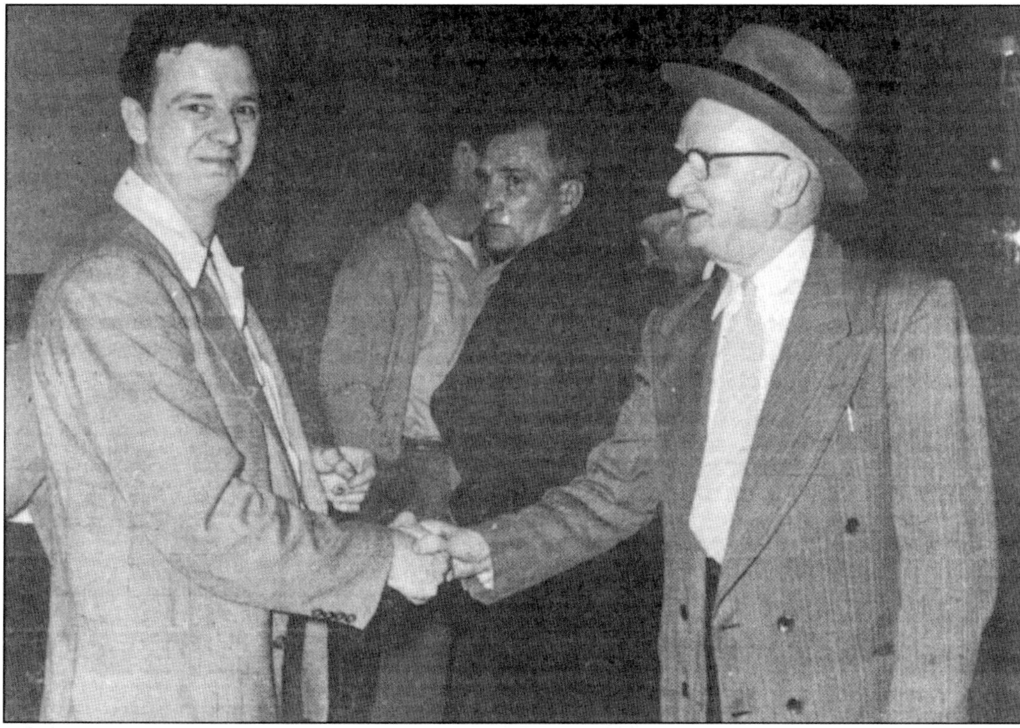

Joe Boyle (left) is congratulated on his successful campaign to unite the Mauch Chunks under the name of Jim Thorpe.

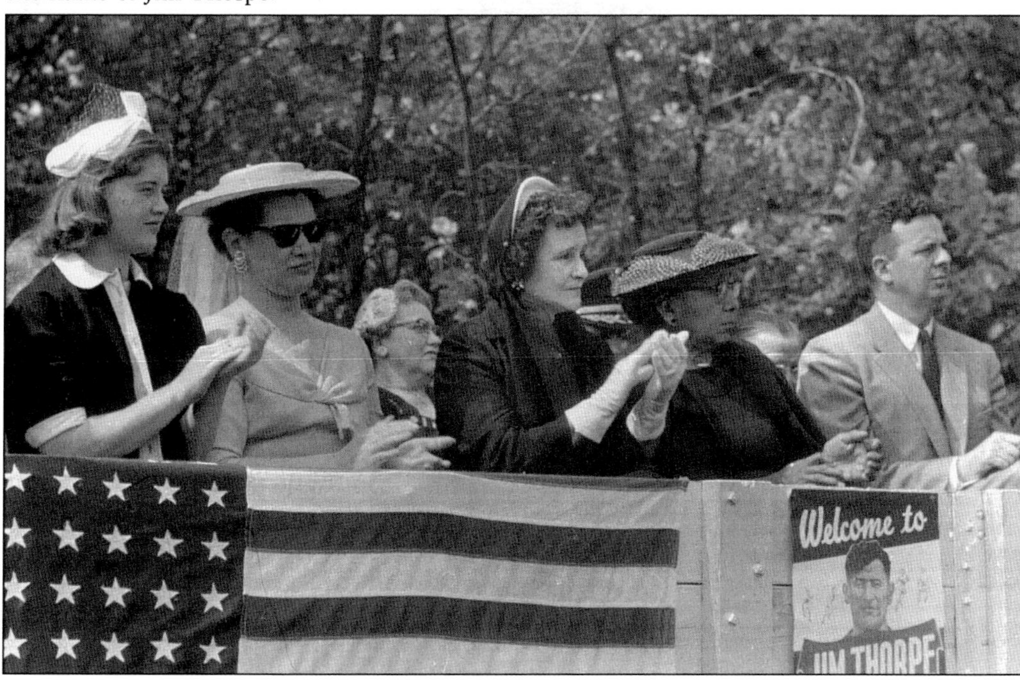

On Memorial Day 1957, Jim Thorpe was laid to rest for the third time in a red granite mausoleum, emblazoned with his athletic achievements, in East Jim Thorpe. Patricia Thorpe (center) is shown with Joe Boyle (right).

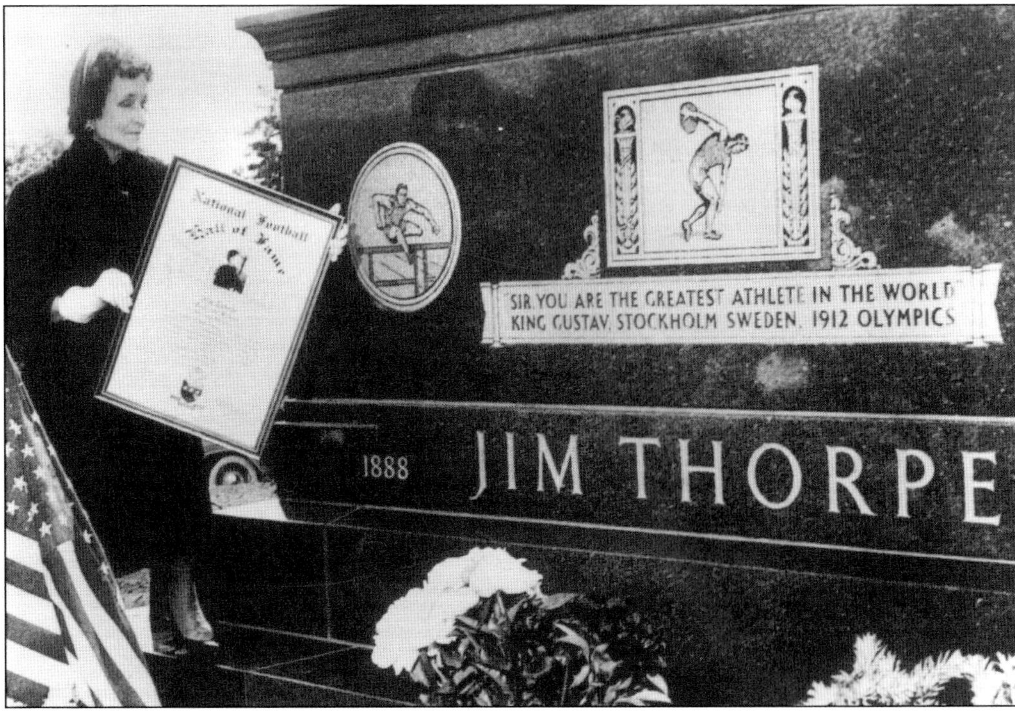

Patricia Thorpe receives a memorial plaque from the citizens of Jim Thorpe. In a telephone interview following the interment, Patricia Thorpe said she was "eternally grateful. . . . Great things will come to pass in the aftermath of this magnificent tribute to my husband."

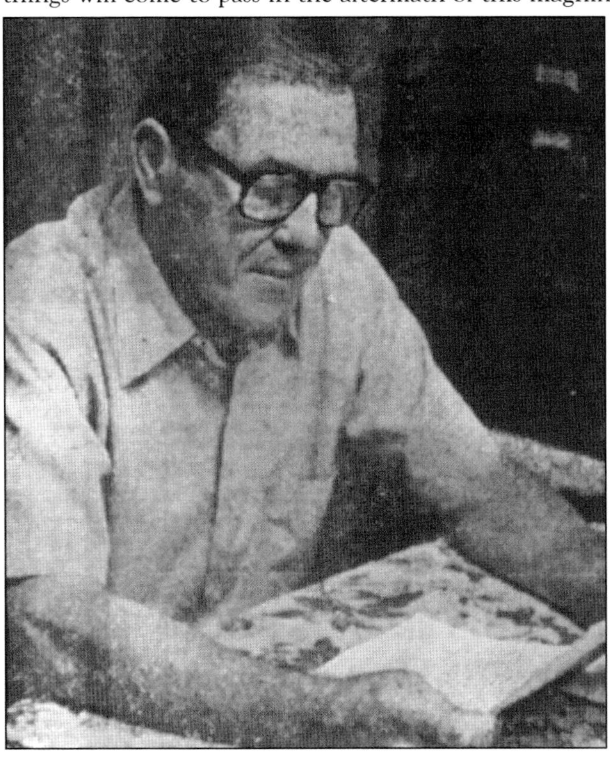

Mayor Michael Hickock looks at letters he received regarding the name change. Controversy did not end with the vote and the burial of Jim Thorpe. Some of the athlete's children opposed their father's body being "sold" by their stepmother as a tourist attraction. Diehard Mauch Chunkers despised the town's new name. Letters from Sac and Fox tribe members and people all over the world demanded the return of Jim Thorpe's body to Oklahoma. Others accused the town of taking advantage of his destitute family.

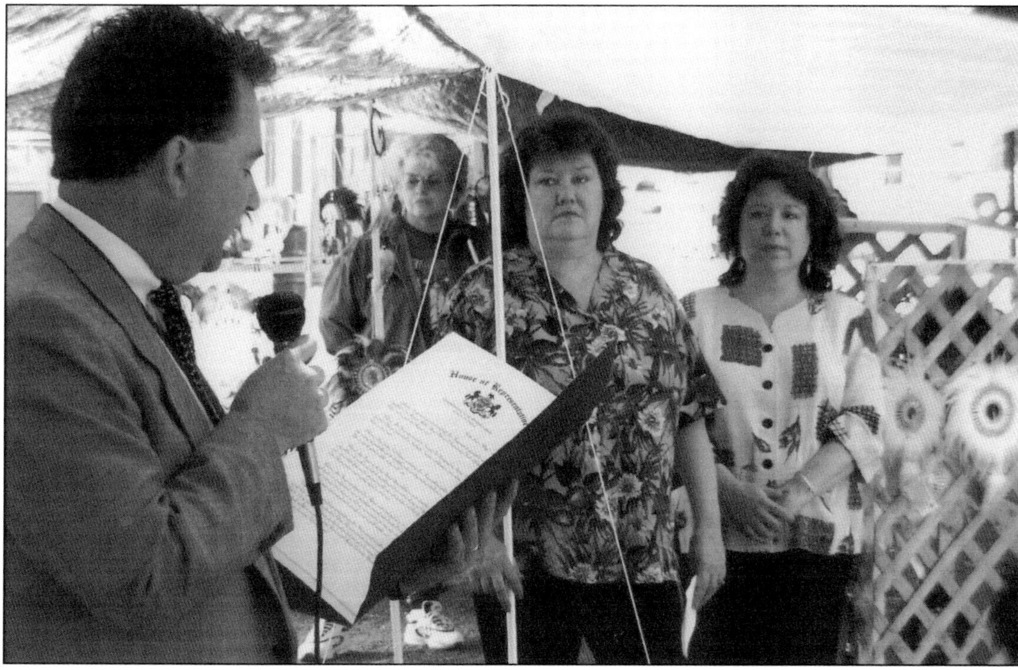

The Thorpe family remains divided about the last resting place of their father. Jim Thorpe's sons want the body returned to Oklahoma. His daughter Grace, however, and her children are often present at ceremonies honoring their father and grandfather. Here, state representative Keith McCall reads a proclamation from the Pennsylvania House of Representatives declaring May 19, 2001, James Francis Thorpe Day in Pennsylvania. Thorpe's granddaughters Teresa Thorpe Gwen and Carla Thorpe look on. (Courtesy Times-News.)

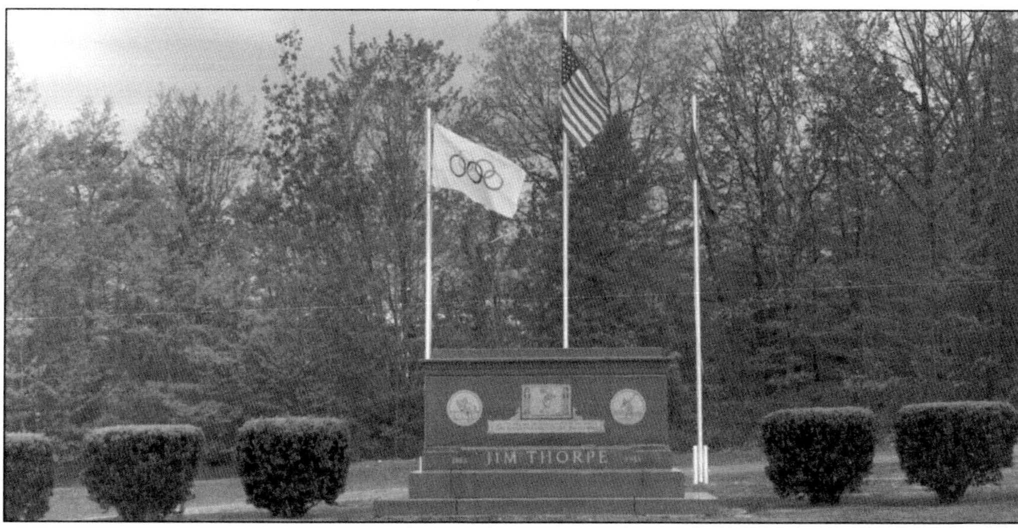

Michael Sofranko, Jim Thorpe Council president, "sees no reason for Thorpe's final resting place to be moved elsewhere. If Jim Thorpe could see what the town has done for him, I think he would be very much pleased." —From the *Times-News*, January 15, 2001.

The majority of townspeople agree. As long as the boroughs of East Mauch Chunk and Mauch Chunk are officially known as Jim Thorpe, it would appear, from the 1954 agreement, that the town will keep its name.

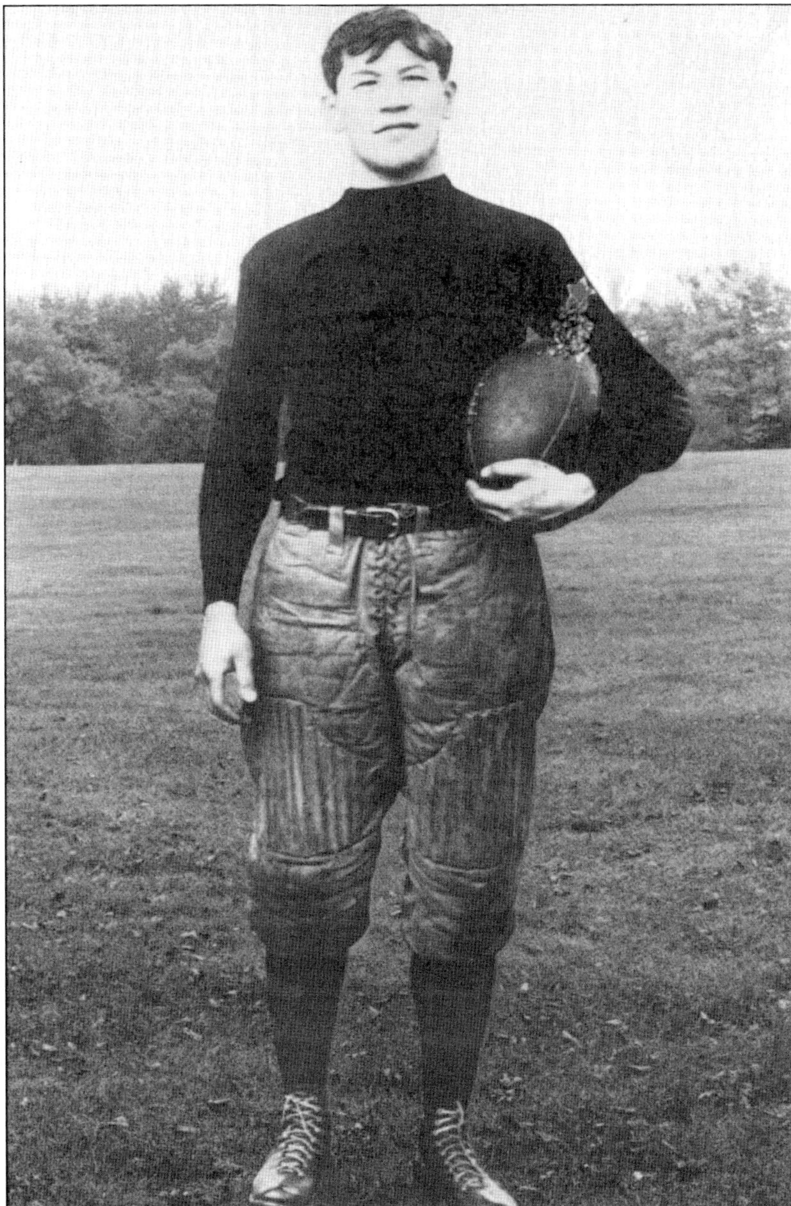

Named Wathohuck (bright spirit), Jim Thorpe was born on the Sac and Fox reservation in Keokuk Falls, Oklahoma, on May 22, 1887. The loss of his twin brother, Charlie, at age eight had a traumatic effect on Jim. He ran away from home and school repeatedly, eventually landing at the Carlisle Indian School to learn a trade. At Carlisle, under the guidance of athletic coach Pop Warner, he became the school's star football player and later went on to become an Olympic hero. In the words of Pop Warner, "he could go skidding through first and second defense; knock off a tackler, stop short and turn past another, ward off still another, and escape the entire pack; or, finally cornered, could go further with a tackler than any man I ever knew." When Carlisle upset West Point in 1912, Dwight D. Eisenhower, a halfback for Army, declared, "Except for Jim Thorpe, Carlisle would have been an easy team to beat. On the football field, there is no one like him in the world."

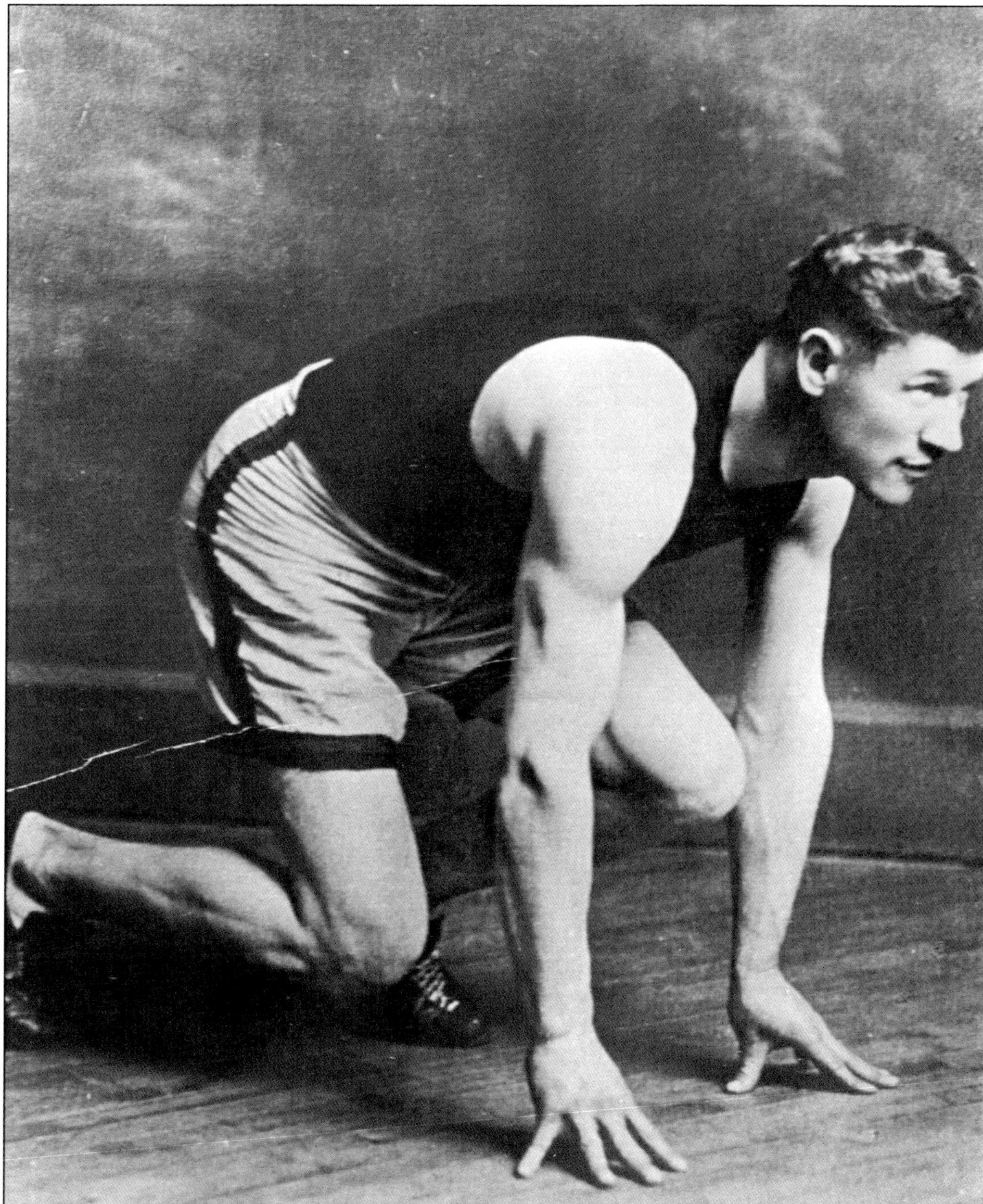

In 1910, Jim Thorpe dropped out of Carlisle Indian School and headed for North Carolina to play semiprofessional baseball for Rocky Mount in the East Carolina League. With an eye on the 1912 Olympics, Pop Warner sent word to him to return. Attracted by Warner's offer to train for the Olympics and play football for Carlisle, Jim Thorpe came back and, on June 14, 1912, set sail for Sweden with the 164-member U.S. Olympic team. Victorious in both the pentathlon and decathlon—a feat that has never been duplicated—the Fifth Olympiad became known as "the Olympics of Jim Thorpe."

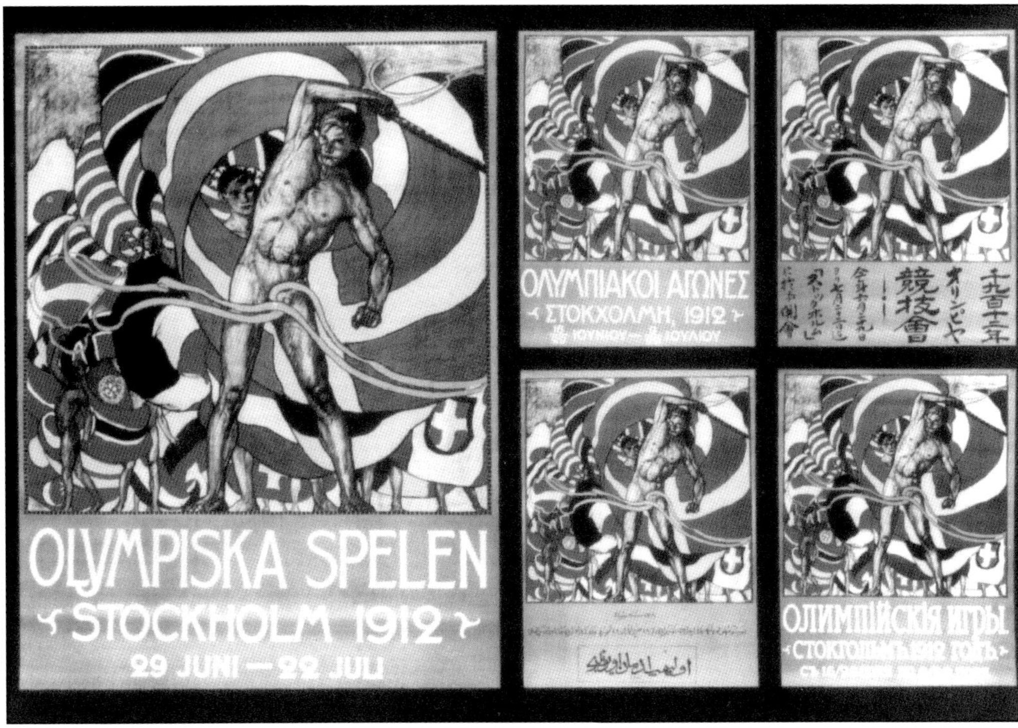

Several years ago, Lars Wass, an official of the Swedish Association for the Promotion of Sports, came into the Mauch Chunk Museum to see the Jim Thorpe exhibit. When he returned to Sweden, he presented the museum with these unique images: the emblem of the 1912 Stockholm Olympics (above) and the Olympic Stadium (below), where Jim Thorpe performed his greatest athletic feats. (Courtesy Lars Wass, Swedish Association for Promotion of Sports.)

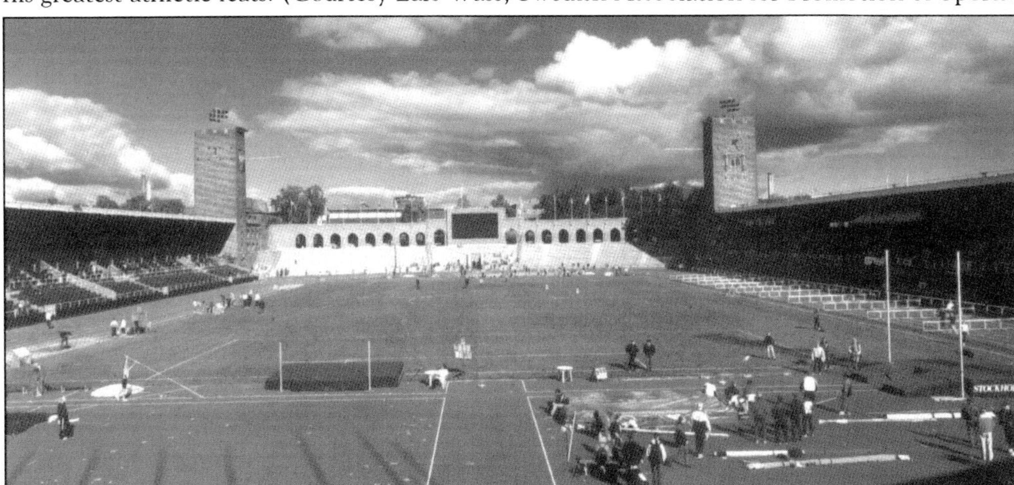

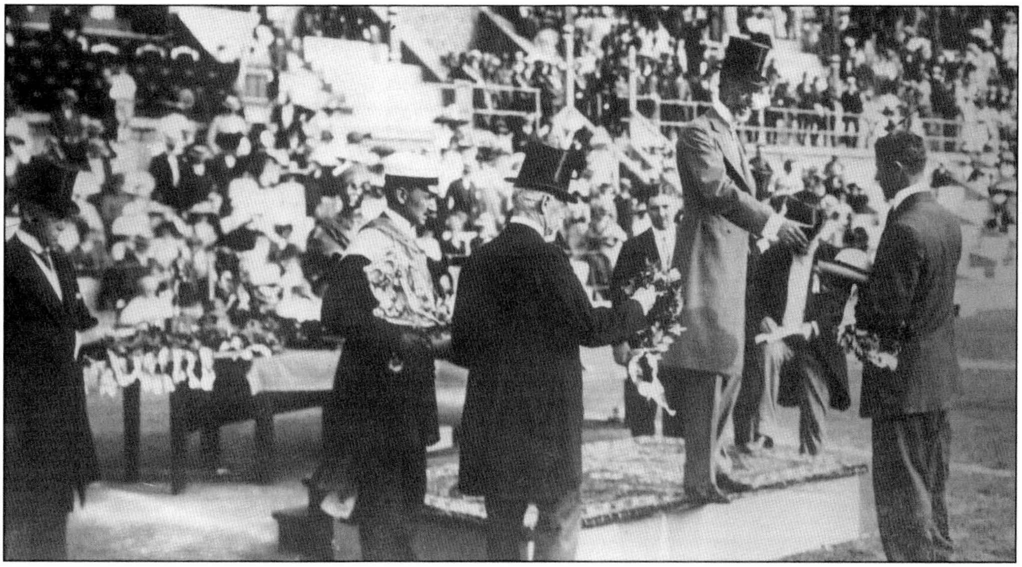

"You, sir, are the greatest athlete in the world!" said King Gustav of Sweden as he presented gold medals to Jim Thorpe. On January 28, 1913, the *New York Times* headline read, "Olympic Prize Lost: Thorpe No Amateur." Jim Thorpe had broken the Olympic rules by playing minor league baseball. As a result, his Olympic records were stricken and his gold medals awarded to pentathlon runner-up, Ferdinand Bie of Norway, and decathlon runner-up, Hugo Weislander of Sweden. Following the Olympics, Jim Thorpe played for eight professional football teams and for the New York Giants baseball team. In 1929, he hung up his cleats. Drinking heavily, he went from job to job, barely surviving. When he passed away of a massive heart attack on March 28, 1953, in California, "the greatest athlete in the world" was destitute. In 1982, his gold medals were restored posthumously.

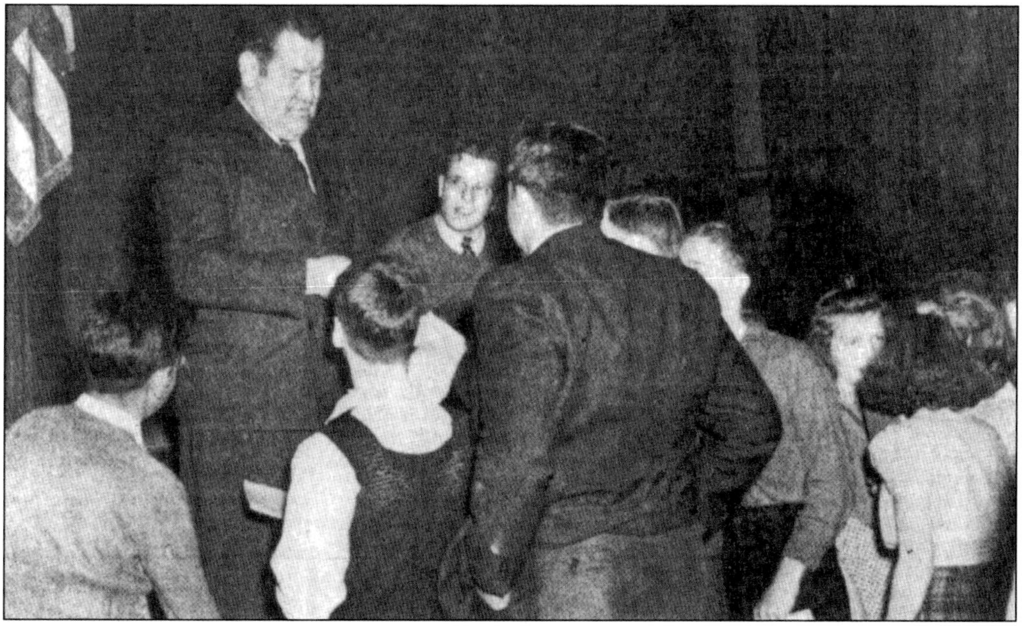

Although Jim Thorpe attended the Carlisle Indian School, his visit to Palmerton was the nearest he came to the town of Jim Thorpe, his namesake and last resting place.

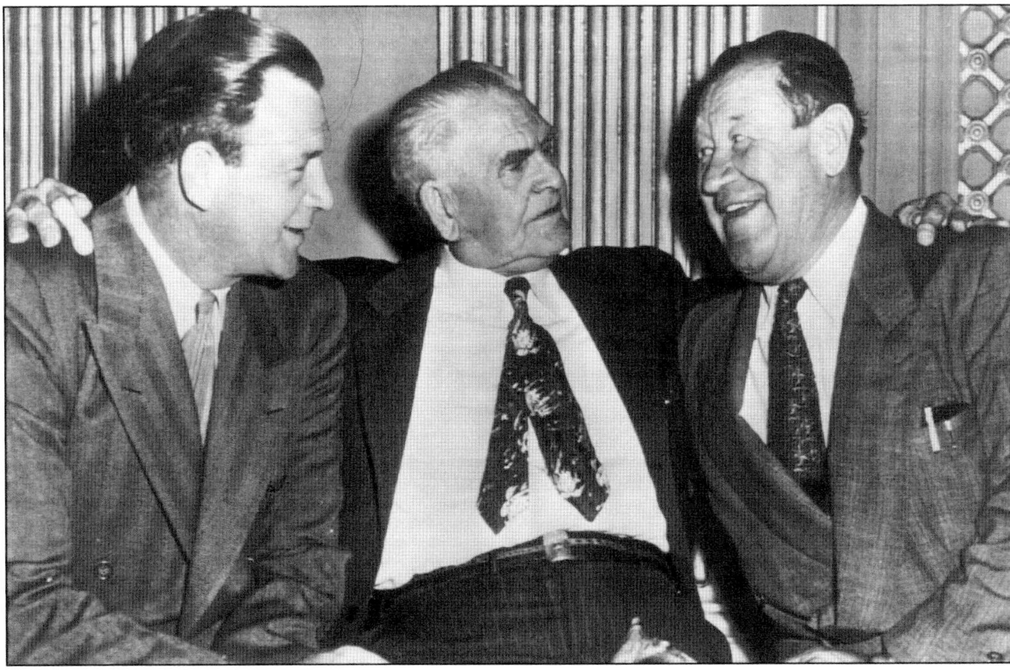

Jim Thorpe gets together with his old Carlisle coach Glen "Pop" Warner and Ernie Nevers in 1948. (Courtesy Smithsonian Institution National Anthropological Archives.)

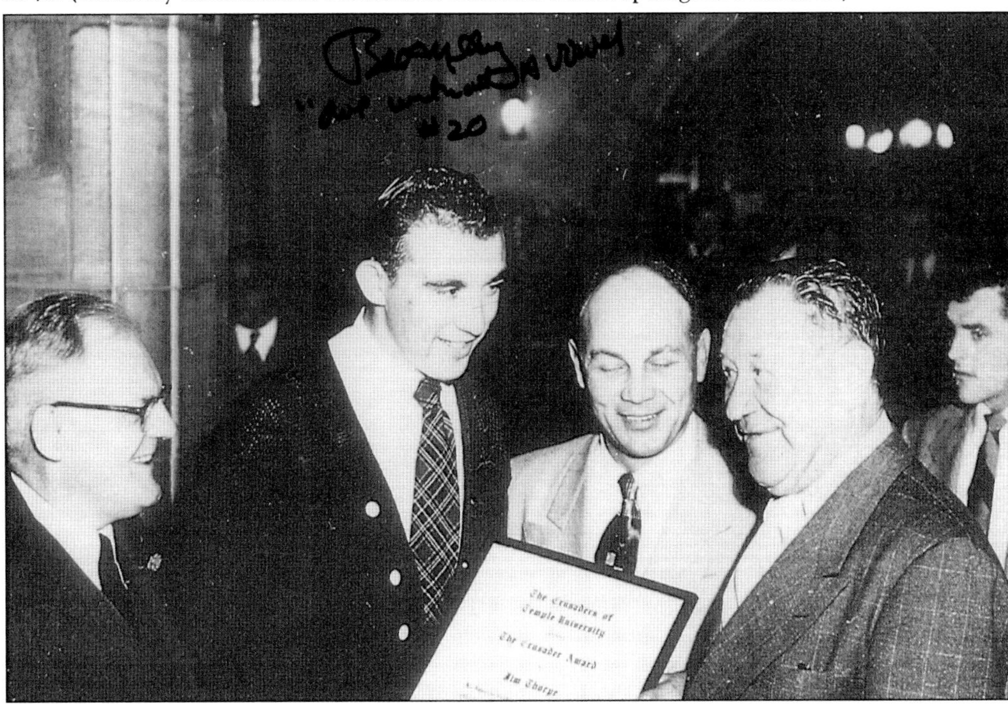

Jim Thorpe receives the Crusader Award from Temple University in Philadelphia in 1952. Shown, from left to right, are Dr. William Tomlinson, vice president of Temple University; William Mlkvy, national basketball champion at Temple University; Al Kawal, football coach; and Jim Thorpe. (Courtesy William Mlkvy.)

Four

Revival

Revitalization did not come with the town's change of name and the burial of Olympic hero Jim Thorpe. It came through the vision of those who saw hope in despair, grandeur in dilapidated buildings, and history in neglected Switchback trails, railroad tracks, the overgrown canal, and the beauty of the Lehigh River and surrounding mountains. Among the visionaries was Agnes McCartney, director of tourism in Carbon County for 30 years.

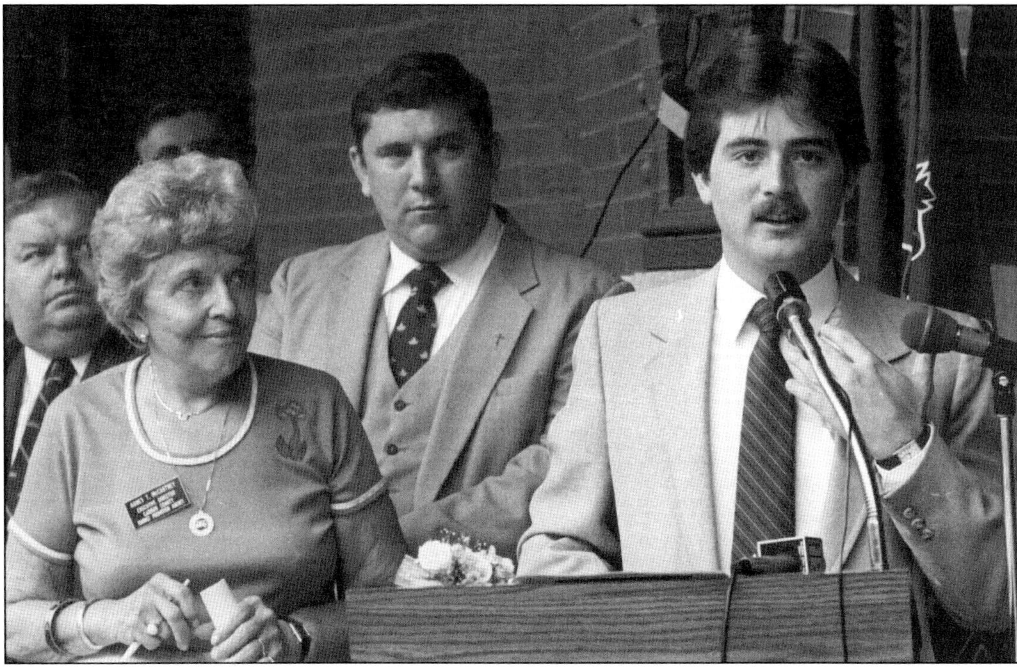

Agnes McCartney introduces politicians outside the tourist promotion headquarters in the former Jersey Central Railroad Station. State representative Keith McCall is at the microphone. Senator Rhoades is in the background. (Courtesy Agnes McCartney.)

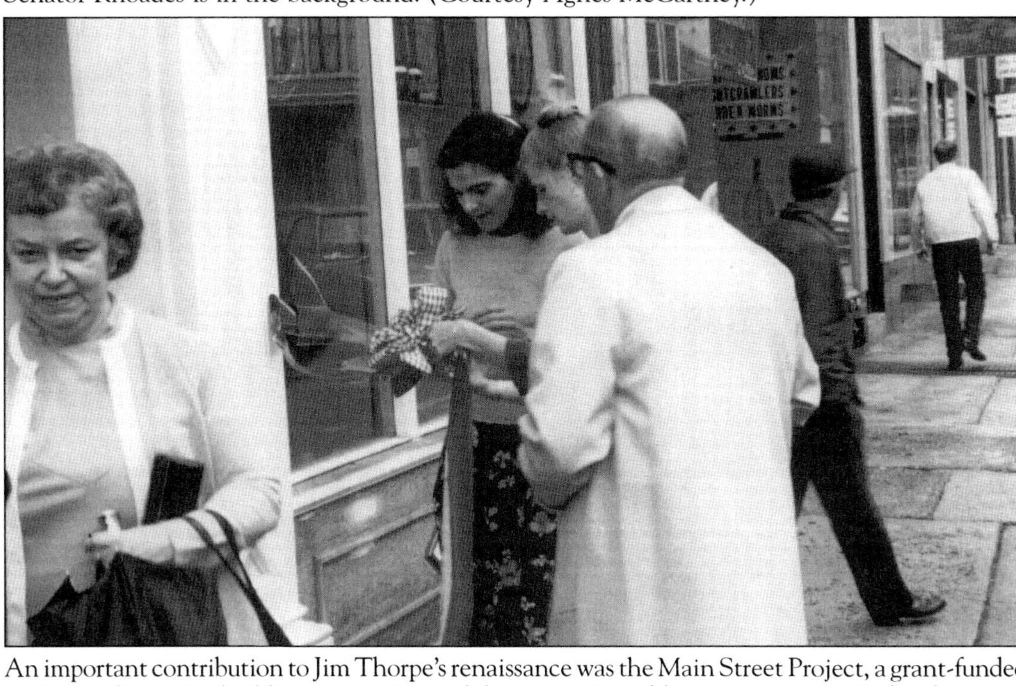

An important contribution to Jim Thorpe's renaissance was the Main Street Project, a grant-funded project underwriting building restoration and the promotion of downtown economic development. Main Street manager Elissa Marsden (left center) assists Dee McClosky with cutting the ribbon at the 1982 grand opening of the Jim Thorpe Arts & Crafts Cooperative in the newly renovated Peter Vathis Building on Lower Broadway. (Courtesy Main Street Archives.)

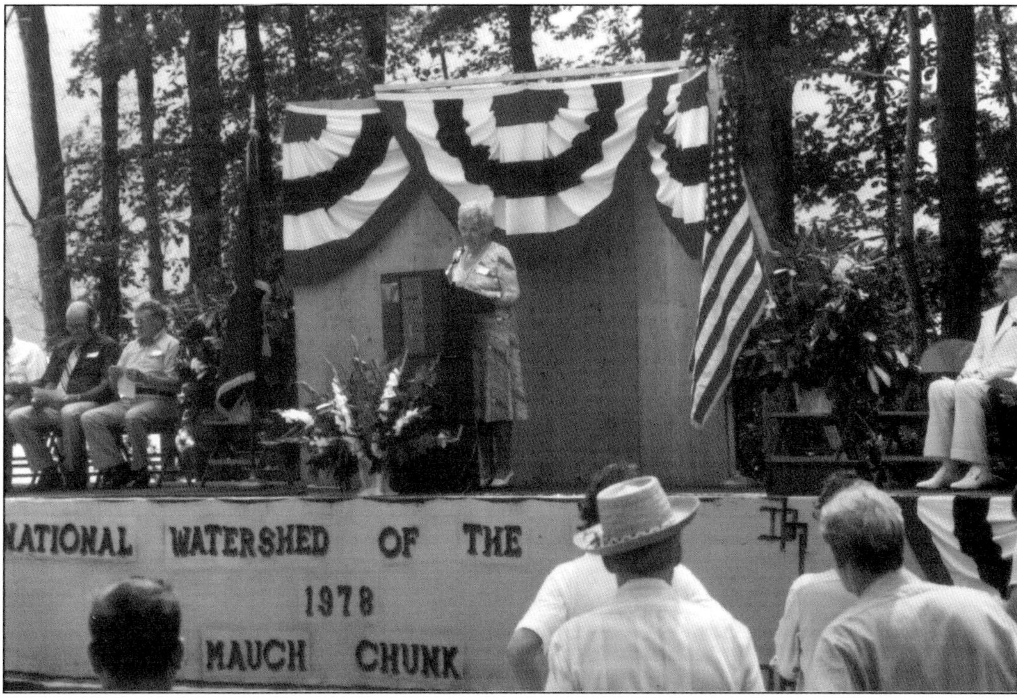

The dedication of Mauch Chunk Creek Dam took place on August 17, 1974, and led to the creation of one of Carbon County's greatest assets, beautiful Mauch Chunk Lake Park. Serving two purposes, the dam controls flooding in Jim Thorpe and provides the town with a 2.5-mile-long, 325-acre lake and park, offering facilities for camping, boating, fishing, swimming, biking, hiking, and cross-country skiing.

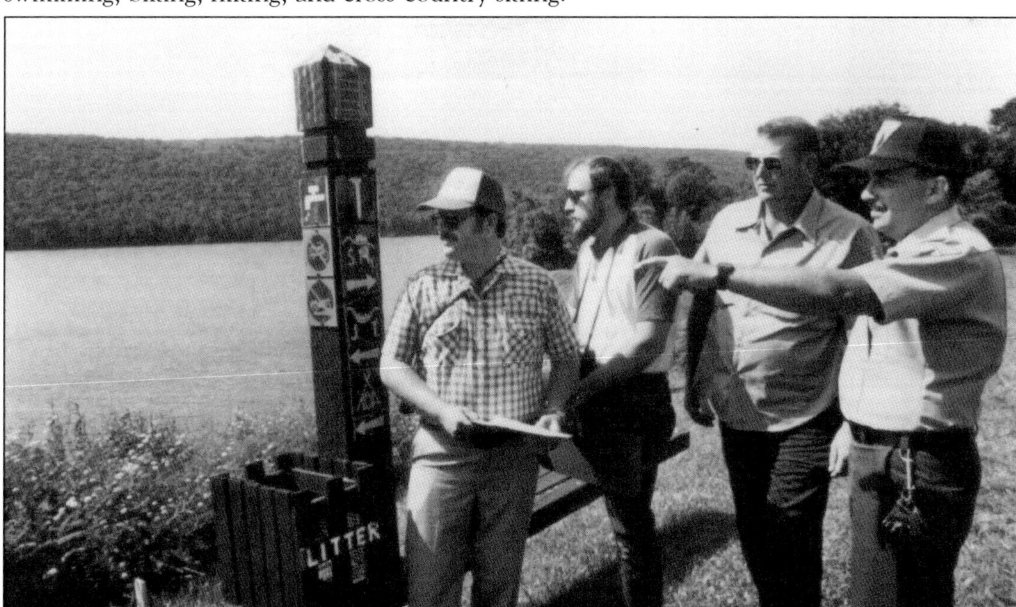

Government engineers inspect the dam with Dennis De Mara (right), Mauch Chunk Lake Park's first director, whose vision and dedication are responsible for today's beautiful park.

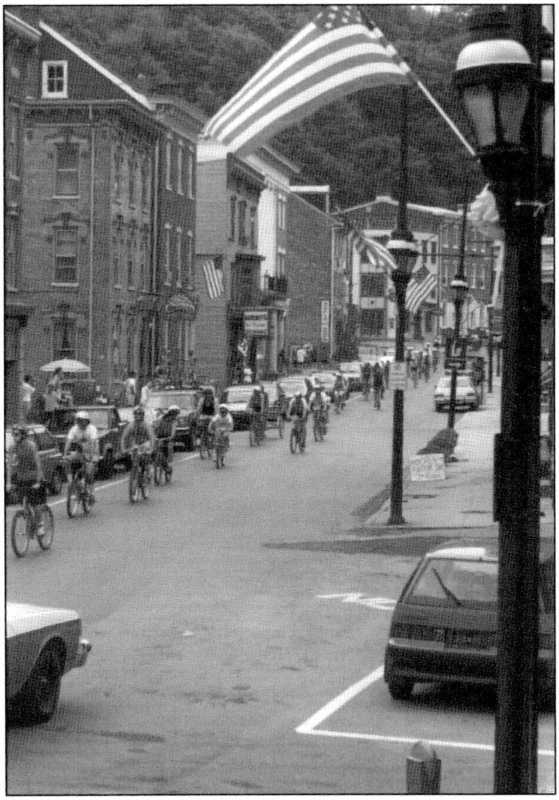

Mountain bikers head out for one of the trails during a fall bike tour. The Upper & Lower Divisions of the Lehigh Canal are links in the 150-mile Delaware & Lehigh National Heritage Corridor, which runs from Bucks County to Luzerne County. A popular haunt of hikers, bikers, and cross-country skiers, the scenic trail follows the Lehigh Canal towpath north through the Narrows, between Flagstaff and Bear Mountains, and then passes along the Jersey Central rail bed north through the Lehigh Gorge to White Haven, the onetime terminus of the Upper Division of the Lehigh Canal. A Corridor Landing (museum and information center), the Pocono Mountain Vacation Bureau, and the Switchback Gravity Railroad Foundation are housed in the former Jersey Central Station. (Courtesy Blue Mountain Sports.)

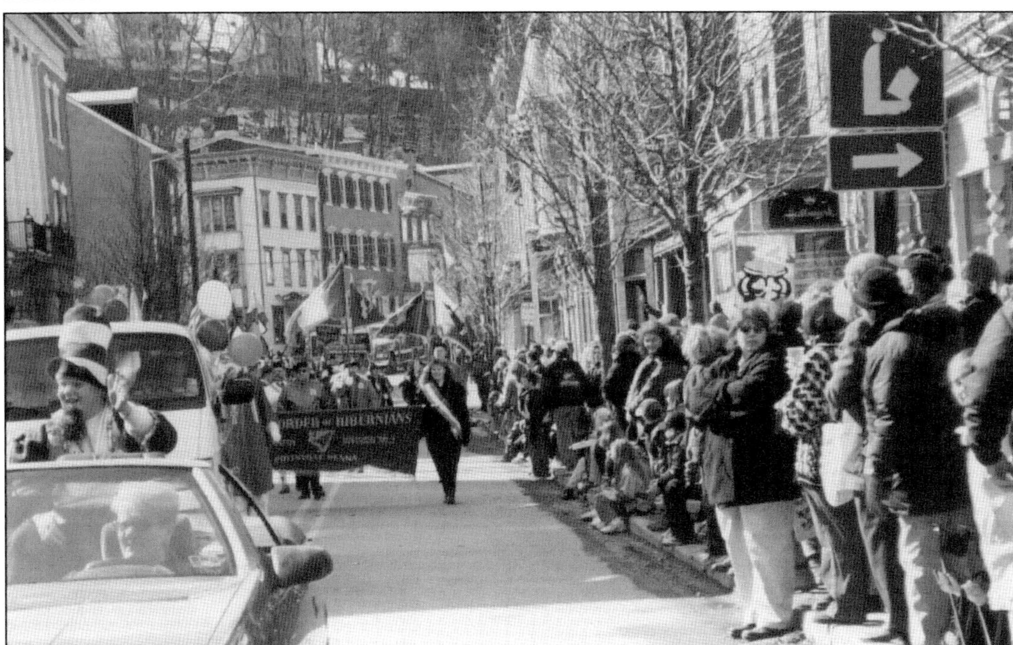

The colorful St. Patrick's Day parade marches down Lower Broadway to the reviewing stand on the balcony of the Inn at Jim Thorpe. (Courtesy Anne Baddick.)

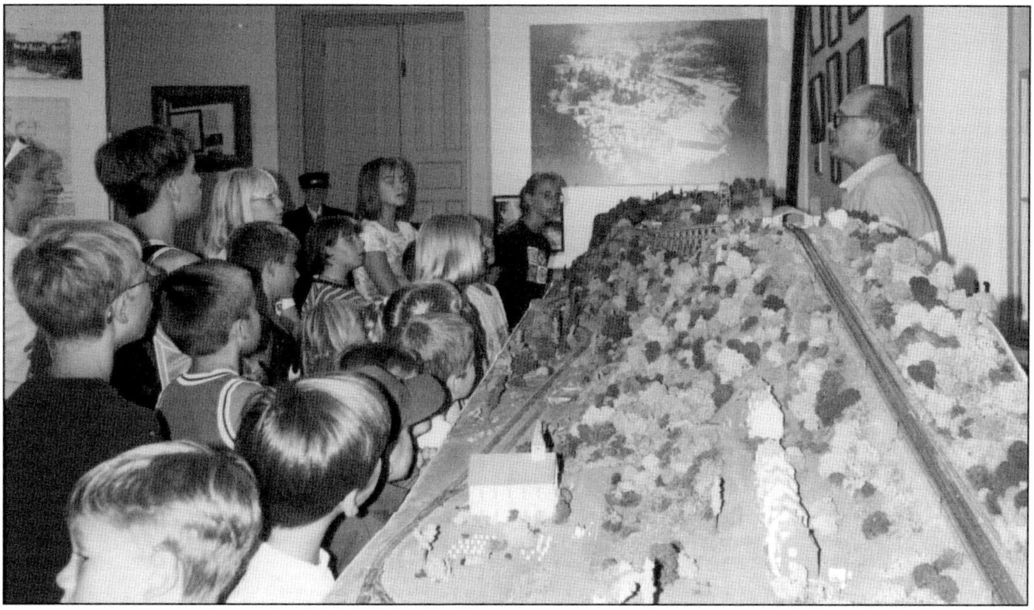

The Switchback rides again at the Mauch Chunk Museum at 41 West Broadway. Located in the former St. Paul's Methodist Church, the Mauch Chunk Museum tells the story of Mauch Chunk. One of the main attractions is the 32-foot working model of the Switchback Gravity Railroad, built by Walter H. Niehoff. Norman Ellis operates the model for schoolchildren and tells them about the history of the Switchback.

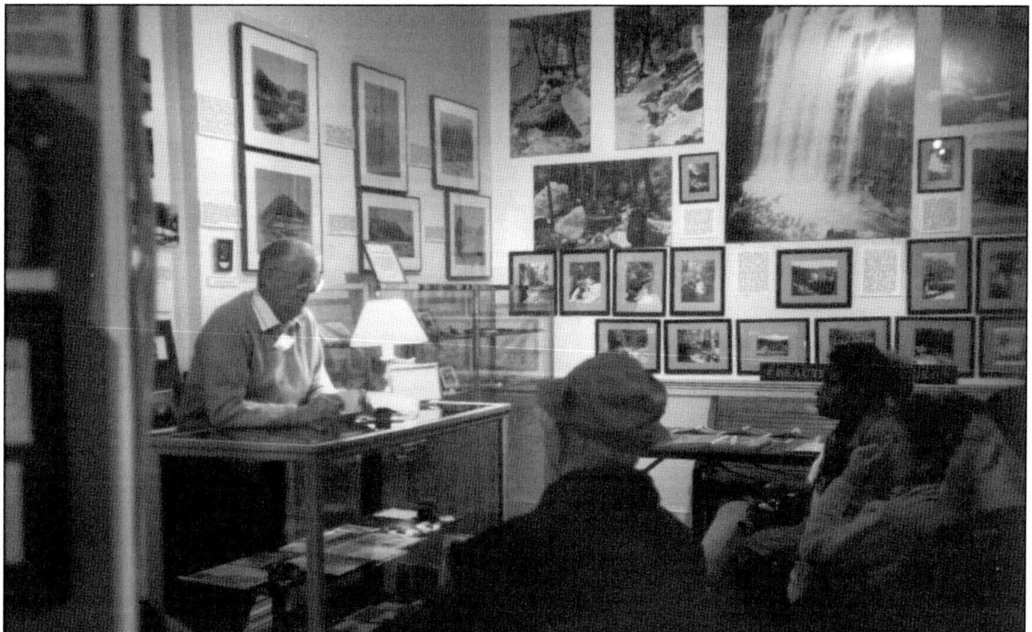

The Elderhostel, sponsored by the Mauch Chunk Museum, visits Jim Thorpe. In the museum, 89-year-old volunteer Tom Potter reminisces about his life in Jim Thorpe. (Courtesy Mary Wohnhaas.)

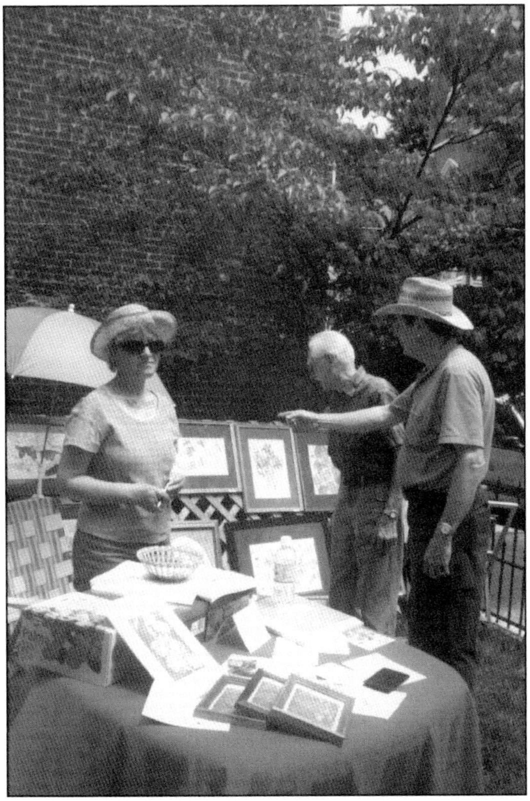

Jean Perry displays her drawings and paintings during Art Walk, 2001. Changing exhibitions of art and sculpture may be viewed at the Maria Feliz Gallery and the Anita Shapolsky Art Foundation on West Broadway.

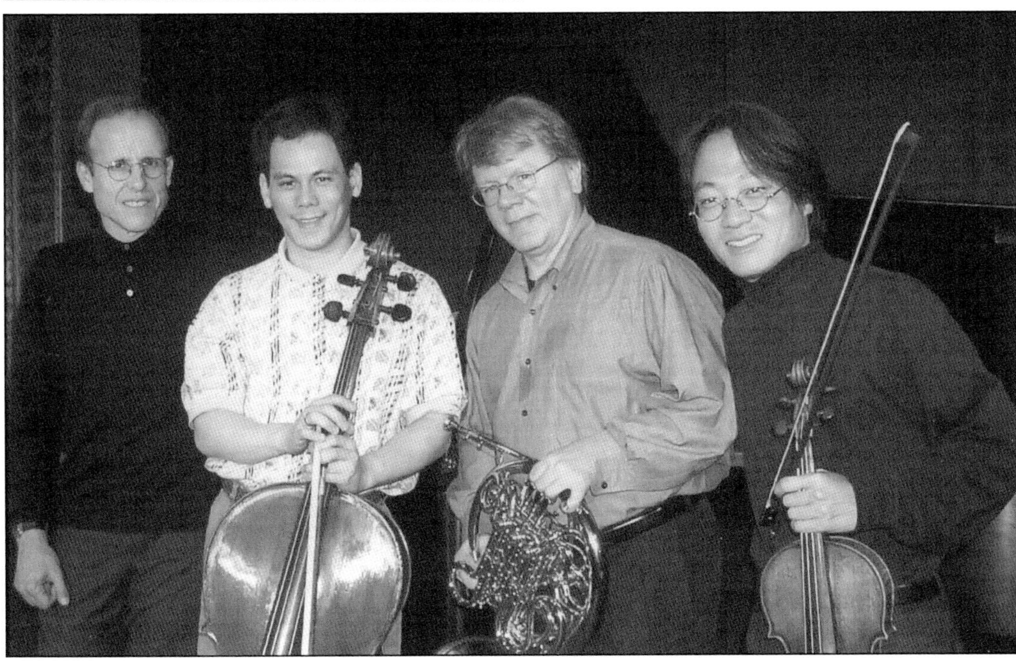

During the Laurel Festival of the Arts, Andre-Michel Schub, Bion Tsang, William Purvis, and Scott Yoo take a break from rehearsing the music of Erwin Schulhoff, Igor Stravinsky, and Johannes Brahms at the Mauch Chunk Opera House. (Courtesy Times-News.)